ENGINEE......

COMPUTER RELAYING FOR POWER SYSTEMS

COMPUTER RELAYING FOR POWER SYSTEMS

Second Edition

Arun G. Phadke

University Distinguished Professor Emeritus
The Bradley Department of Electrical and Computer Engineering
Virginia Tech, Blacksburg, Virginia, USA

James S. Thorp

Hugh P. and Ethel C. Kelley Professor and Department Head
The Bradley Department of Electrical and Computer Engineering
Virginia Tech, Blacksburg, Virginia, USA

A John Wiley and Sons, Ltd., Publication

Research Studies Press Limited

Copyright © 2009   Research Studies Press Limited, 16 Coach House Cloisters, 10 Hitchin Street, Baldock,
Hertfordshire, SG7 6AE

Published by   John Wiley & Sons Ltd, The Atrium, Southern Gate, Chichester,
West Sussex PO19 8SQ, England
Telephone (+44) 1243 779777

Email (for orders and customer service enquiries): cs-books@wiley.co.uk
Visit our Home Page on www.wileyeurope.com or www.wiley.com

This Work is a co-publication between Research Studies Press Limited and John Wiley & Sons, Ltd.

This edition first published 2009

MATLAB® MATLAB and any associated trademarks used in this book are the registered trademarks of The
MathWorks, Inc.

Library of Congress Cataloging-in-Publication Data:

Phadke, Arun G.
    Computer relaying for power systems / Arun G. Phadke. – 2nd ed.
        p. cm.
    Includes bibliographical references and index.
    ISBN 978-0-470-05713-1 (cloth)
1. Protective relays. 2. Electric power systems – Protection – Data processing. I. Title.
    TK2861.P48 2009
    621.31'7 – dc22
                                                    2009022672

A catalogue record for this book is available from the British Library.

ISBN 978-0-470-05713-1 (H/B)

Typeset in 11/13 Times by Laserwords Private Limited, Chennai, India.
Printed and bound in Great Britain by CPI Antony Rowe, Chippenham, Wiltshire

CONTENTS

About the authors

Dr. Arun G. Phadke worked in the Electric Utility industry for 13 years before joining Virginia Tech 1982. He became the American Electric Power Professor of Electrical Engineering in 1985 and held this title until 2000 when he was recognized as a University Distinguished Professor. He became University Distinguished Professor Emeritus in 2003, and continues as a Research Faculty member of the Electrical and Computer Engineering Department of Virginia Tech. Dr. Phadke was elected a Fellow of IEEE in 1980. He was elected to the National Academy of Engineering in 1993. He was Editor in Chief of Transactions of IEEE on Power Delivery. He became the Chairman of the Power System Relaying Committee of IEEE in 1999–2000. Dr. Phadke received the Herman Halperin award of IEEE in 2000. Dr. Phadke has also been very active in CIGRE. He has been a member of the Executive Committee of the US National Committee of CIGRE, and was the Chairman of their Technical Committee. He was previously the Vice President of USNC-CIGRE and served as Secretary/Treasurer. In 2002 he was elected a 'Distinguished Member of CIGRE' by the Governing Board of CIGRE. Dr. Phadke was active in CIGRE SC34 for several years, and was the Chairman of some of their working groups. In 1999 Dr. Phadke joined colleagues from Europe and Far East in founding the International Institute for Critical Infrastructures (CRIS). He was the first President of CRIS from 1999–2002, and currently serves on its Governing Board. Dr. Phadke received the 'Docteur Honoris Causa' from *Institute National Polytechnic de Grenoble (INPG)* in 2006. Dr. Phadke received the 'Karapetoff Award' from the HKN Society, and the 'Benjamin Franklin Medal' for Electrical Engineering in 2008.

Dr. James S. Thorp is the Hugh P. and Ethel C. Kelley Professor of Electrical and Computer Engineering and Department Head of the Bradley Department of Electrical and Computer Engineering at Virginia Tech. He was the Charles N. Mellowes Professor in Engineering at Cornell University from 1994–2004. He obtained the B.E.E. in 1959 and the Ph. D. in 1962 from Cornell University and was the Director of the Cornell School of Electrical and Computer Engineering from 1994 to 2001, a Faculty Intern, American Electric Power Service Corporation in 1976–77 and an Overseas Fellow, Churchill College, Cambridge University in 1988. He has consulted for Mehta Tech Inc., Basler Electric, RFL Dowty Industries,

American Electric Power Service Corporation, and General Electric. He was an Alfred P. Sloan Foundation National Scholar and was elected a Fellow of the IEEE in 1989 and a Member of the National Academy of Engineering in 1996. He received the 2001 Power Engineering Society Career Service award, the 2006 IEEE Outstanding Power Engineering Educator Award, and shared the 2007 Benjamin Franklin Medal with A.G. Phadke.

Preface to the first edition

The concept of using digital computers for relaying originated some 25 years ago. Since then the field has grown rapidly. Computers have undergone a significant change – they have become more powerful, cheaper, and sturdier. Today computer relays are preferred for economic as well as technical reasons. These advances in computer hardware have been accompanied by analytical developments in the field of relaying. Through the participation of researchers at Universities and industrial organizations, the theory of power system protection has been placed on a mathematical basis. It is noted that, in most cases, the mathematical investigations have confirmed the fact that traditional relay designs have been optimum or near-optimum solutions to the relaying problem. This is reassuring: the theory and practice of relaying have been reaffirmed simultaneously.

An account of these developments is scattered throughout the technical literature: Proceedings of various conferences, Transactions of Engineering Societies, and technical publications of various equipment manufacturers. This book is our attempt to present a coherent account of the field of computer relaying. We have been doing active research in this area – much of it in close collaboration with each other – since the mid 1970's. We have tried to present a balanced view of all the developments in the field, although it may seem that, at times, we have given a fuller account of areas in which we ourselves have made contributions. For this bias – if it is perceived as such by the reader – we seek his indulgence.

The book is intended for graduate students in electric power engineering, for researchers in the field, or for anyone who wishes to understand this new development in the role of a potential user or manufacturer of computer relays. In teaching a course from this book, we recommend following the order of the material in the book. If a course on traditional protection is a pre-requisite to this course, Chapter 2 may be omitted. The mathematical basis for relaying is contained in Chapter 3, and is intended for those who are not in an academic environment at present. The material is essential for gaining an understanding of the reason *why* a relaying algorithm works as it does, although *how* an algorithm works – i.e. its procedural structure – can be understood without a thorough knowledge of the mathematics. A reader with such a limited objective may skip the mathematical background, and go directly to the sections of immediate interest to him.

Our long association with the American Electric Power Service Corporation (AEP) has been the single most important element in sustaining our interest in Computer Relaying. The atmosphere in the old Computer Applications Department in AEP under Tony Gabrielle was particularly well suited for innovative engineering. He was responsible for starting us on this subject, and for giving much needed support when practical results seemed to be far into the future. Also present at AEP was Stan Horowitz, our colleague and teacher, without whose help we would have lost touch with the reality of relaying as a practical engineering enterprise. Stan Horowitz, Eric Udren, and Peter McLaren read through the manuscript and offered many constructive comments. We are grateful for their help. The responsibility for the book, and for any remaining errors, is of course our own.

We continue to derive great pleasure from working in this field. It is our hope that, with this book, we may share this enjoyment with the reader.

Arun G. Phadke
Blacksburg

James S. Thorp
Ithaca
1988

Preface to the second edition

The first edition of this book was published in 1988. The intervening two decades have seen wide-spread acceptance of computer relays by power engineers through-out the world. In fact, in many countries computer relays are the protective devices of choice, and one would be hard pressed to find electromechanical or electronic relays with comparable capabilities. Clearly economics of relay manufacture have played a major role in making this possible, and the improved performance, self-checking capabilities, and access to relay settings over communication lines have been the principal features of this technology which have brought about their acceptance on such a wide scale.

It has been recognized by most relay designers – and is also the belief of the authors – that the principles of protection have essentially remained as established by experience gained over the last century. Computer relays provide essentially the same capabilities as traditional relays in a more efficient manner. Having said this, it is also recognized that changes in protection principles have taken place, solely because of the capabilities of the computers and the available communication facilities. Thus adaptive relaying could not be realized without this new technology. Adaptive relaying, along with the new field of wide area measurements (which originated in the field of computer relaying) forms a significant part of the present edition of our book.

A study of published research papers on relaying will show that researchers continue to investigate the application of newer analytical techniques to the field of relaying. We have included an account of several such techniques in this edition, but it must be stated that most of these techniques have not seen their implementation in practical relay designs. Perhaps this confirms the authors' belief that the principles of protection are essentially dictated by power system phenomena, and the long established techniques of protection system design are very sound and close to being optimum. The newer analytical techniques which are being investigated offer very minor improvements at best, and it remains questionable as to when or for which applications we will see a clear benefit of these newer analytical techniques.

Our book remains a research text and reference work. As such the problem set at the end of each chapter is often a statement of research idea. Some problems are quite complex, and each problem leaves room for individual interpretation and development. We therefore offer no solutions to these problems and leave their resolution to the individual initiative of the reader. We are of course interested in receiving any comments that the users of our book care to make.

The authors have participated with pleasure in project "111", a Key Research Project of the North China Electric Power University since its inception in 2008 under the direction of Professor Yang Qixun. In addition to promoting research in many aspects of computer relaying in which the authors continue to participate, the facilities provided in Beijing under the auspices of this project for the authors have facilitated the timely completion of this Second Edition of our book.

We continue to derive great pleasure from working in this field. It is our hope that, with the second edition of this book, we may share this enjoyment with the reader.

Arun G. Phadke
Blacksburg

James S. Thorp
Blacksburg
2009

Glossary of acronyms

A/D	Analog to Digital
ADC	Analog to Digital Converter
ANN	Artificial Neural Network
ANSI	American National Standards Institute
CIGRE	International Council on Large Electric Systems
CT	Current Transformer
CVT	Capacitive Voltage Transformer
DFT	Discrete Fourier Transform
EHV	Extra High Voltage
EMI	Electromagnetic Interference
EMTP	Electromagnetic Transients Program
EPRI	Electric Power Research Institute
EPROM	Erasable Programmable Read Only Memory
FFT	Fast Fourier Transform
GPS	Global Positioning System
I/O	Input Output
IEC	International Electrotechnical Commission
IEEE	Institute of Electronic and Electrical Engineers
MOV	Metal Oxide Varistors
MUX	Multiplexer
NAVSTAR	NAVSTAR is not an acronym. It represents GPS described above.
PDC	Phasor Data Concentrator
PMU	Phasor Measurement Unit
PROM	Programmable Read Only Memory
PT	Potential Transformer
RAM	Random Access Memory
RAS	Remedial Action Scheme

ROM	Read Only Memory
S/H	Sample and Hold
SCDFT	Symmetrical Component Discrete Fourier Transform
SIPS	System Integrity Protection Scheme
SWC	Surge Withstand Capability
WAMS	Wide Area Measurement System
WAMPACS	Wide Area Measurement, Protection and Control System
WLS	Weighted Least Squares

1

Introduction to computer relaying

1.1 Development of computer relaying

The field of computer relaying started with attempts to investigate whether power system relaying functions could be performed with a digital computer. These investigations began in the 1960s, a period during which the digital computer was slowly and systematically replacing many of the traditional tools of analytical electric power engineering. The short circuit, load flow, and stability problems – whose solution was the primary preoccupation of power system planners – had already been converted to computer programs, replacing the DC boards and the Network Analyzers. Relaying was thought to be the next promising and exciting field for computerization. It was clear from the outset that digital computers of that period could not handle the technical needs of high speed relaying functions. Nor was there any economic incentive to do so. Computers were orders of magnitude too expensive. Yet, the prospect of developing and examining relaying algorithms looked attractive to several researchers. Through such essentially academic curiosity this very fertile field was initiated. The evolution of computers over the intervening years has been so rapid that algorithmic sophistication demanded by the relaying programs has finally found a correspondence in the speed and economy of the modern microcomputer; so that at present computer relays offer the best economic and technical solution to the protection problems – in many instances the only workable solution. Indeed, we are at the start of an era in which computer relaying has become routine, and it has further influenced the development of effective tools for real-time monitoring and control of power systems.

In this chapter we will briefly review the historical developments in the field of computer relaying. We will then describe the architecture of a typical computer based relay. We will also identify the critical hardware components, and discuss the influence they have on the relaying tasks.

Computer Relaying for Power Systems 2e   by A. G. Phadke and J. S. Thorp
© 2009 John Wiley & Sons, Ltd

1.2   Historical background

One of the earliest published papers on computer relaying explored the somewhat curious idea that relaying of all the equipment in a substation would be handled by a single computer.[1] No doubt this was motivated by the fact that computers were very expensive at that time (1960s), and there could be no conceivable way in which multiple computers would be economically palatable as a substitute for conventional relays which were at least one order of magnitude less expensive than a suitable computer. In addition, the computation speed of contemporary computers was too slow to handle high speed relaying, while the power consumption of the computers was too high. In spite of these obvious shortcomings – which reflected the then current state of computer development – the reference cited above explored several protection algorithmic details thoroughly, and even today provides a good initiation to the novice in the complexities of modern relaying practices.

Several other papers were published at approximately the same time, and led to the algorithmic development for protection of high voltage transmission lines.[2,3] It was recognized early that transmission line protection function (distance relaying in particular) – more than any other – is of greatest interest to relay engineers because of its widespread use on power systems, its relatively high cost, and its functional complexity. These early researchers began a study of distance protection algorithms which continues unabated to this day. These studies have led to important new insights into the physical nature of protection processes and the limits to which they can be pushed. It is quite possible that distance relaying implementation on computers has been mastered by most researchers by now, and that any new advances in this field are likely to come from the use of improved computer hardware to implement the well-understood distance relaying algorithms.

An entirely different approach to distance relaying has been proposed during recent years.[4,5] It is generally based upon the utilization of traveling waves initiated by a fault to estimate the fault distance. Traveling wave relays require relatively high frequencies for sampling voltage and current input signals. Although traveling wave relays have not offered compelling advantages over other relaying principles in terms of speed and accuracy of performance, they have been applied in a few instances around the world with satisfactory performance. This technique will be covered more fully in Chapter 9; it remains for the present a somewhat infrequently used relaying application. Fault location algorithms based on traveling waves have also been developed and there are reports of good experience with these devices. These too will be covered more fully in Chapter 9.

In addition to the development of distance relaying algorithms, work was begun early on apparatus protection using the differential relaying principle.[6–8] These early references recognize the fact that compared to the line relaying task, differential relaying algorithms are less demanding of computational power. Harmonic restraint function adds some complexity to the transformer protection problem, and problems associated with current transformer saturation or other inaccuracies continue to have no easy solutions in computer based protection systems just as in conventional relays. Nevertheless, with the algorithmic development of distance and differential relaying principles, one could say that the ability of computer based relays to provide performance at least as good as conventional relays had been established by the early 1970s.

Very significant advances in computer hardware had taken place since those early days. The size, power consumption, and cost of computers had gone down by orders of magnitude, while simultaneously the speed of computation increased by several orders. The appearance of 16 bit (and more recently of 32 bit) microprocessors and computers based upon them made high speed computer relaying technically achievable, while at the same time cost of computer based relays began to become comparable to that of conventional relays. This trend has continued to the present day – and is bound to persist in the future – although perhaps at not quite as precipitous a rate. In fact, it appears well established by now that the most economical and technically superior way to build relay systems of the future (except possibly for some functionally simple and inexpensive relays) is with digital computers. The old idea of combining several protection functions in one hardware system[1] has also re-emerged to a certain extent – in the present day multi-function relays.

With reasonable prospects of having affordable computer relays which can be dedicated to a single protection function, attention soon turned to the opportunities offered by computer relays to integrate them into a substation-wide, perhaps even a system-wide, network using high-speed wide-band communication networks. Early papers on this subject realized several benefits that would flow from this ability of relays to communicate.[9,10] As will be seen in Chapters 8 and 9 integrated computer systems for substations which handle relaying, monitoring, and control tasks offer novel opportunities for improving overall system performance by exchanging critical information between different devices.

1.3   Expected benefits of computer relaying

It would be well to summarize the advantages offered by computer relays, and some of the features of this technology which have required new operational considerations. Among the benefits flowing from computer relays are:

1.3.1   Cost

All other things being equal, the cost of a relay is the main consideration in its acceptability. In the early stages of computer relaying, computer relay costs were 10 to 20 times greater than the cost of conventional relays. Over the years, the cost of digital computers has steadily declined; at the same time their computational power (measured by instruction execution time and word length) has increased substantially. The cost of conventional (analog) relays has steadily increased over the same period, primarily because of some design improvements, but also because of general inflation and a relatively low volume of production and sales. It is estimated that for equal performance the cost of the most sophisticated digital computer relays (including software costs) would be about the same as that of conventional relaying systems. Clearly there are some conventional relays – overcurrent relays are an example – which are so inexpensive that cheaper computer relays to replace them seem unlikely at present, unless they are a part of a multi-function relay. However, for major protection systems, the competitive computer relay costs have definitely become an important consideration.

1.3.2   Self-checking and reliability

A computer relay can be programmed to monitor several of its hardware and software subsystems continuously, thus detecting any malfunctions that may occur. It can be designed to fail in a safe mode – i.e. take itself out of service if a failure is detected – and send a service request alarm to the system center. This feature of computer relays is perhaps the most telling technical argument in favor of computer relaying. Misoperation of relays is not a frequent occurrence, considering the very large number of relays in existence on a power system. On the other hand, in most cases of power system catastrophic failures the immediate cause of the escalation of events that leads to the failure can be traced to relay misoperation. In some cases, it is a mis-application of a relay to the given protection task, but in a majority of cases it is due to a failure of a relay component that leads to its misoperation and the consequent power system breakdown.[11] It is expected that with the self-checking feature of computer based relays, the relay component failures can be detected soon after they occur, and could be repaired before they have a chance to misoperate. In this sense, although computer based relays are more complex than electromechanical or solid state relays (and hence potentially more likely to fail), as a system they have a higher rate of availability. Of course, a relay cannot detect all component failures – especially those outside the periphery of the relay system.

1.3.3   System integration and digital environment

Digital computers and digital technology have become the basis of most systems in substations. Measurements, communication, telemetry and control are all computer based functions. Many of the power transducers (current and voltage transformers) are in the process of becoming digital systems. Fiber optic links, because of their immunity to Electromagnetic Interference (EMI), are likely to become the medium of signal transmission from one point to another in a substation; it is a technology particularly suited to the digital environment. In substations of the future, computer relays will fit in very naturally. They can accept digital signals obtained from newer transducers and fiber optic channels, and become integrated with the computer based control and monitoring systems of a substation. As a matter of fact, without computer relaying, the digital transducers and fiber optic links for signal transmission would not be viable systems in the substation.

1.3.4   Functional flexibility and adaptive relaying

Since the digital computer can be programmed to perform several functions as long as it has the input and output signals needed for those functions, it is a simple matter to the relay computer to do many other substation tasks. For example, measuring and monitoring flows and voltages in transformers and transmission lines, controlling the opening and closing of circuit breakers and switches, providing backup for other devices that have failed, are all functions that can be taken over by the relay computer. The relaying function calls for intensive computational activity when a fault occurs on the system. This intense activity at best occupies the relaying computer for a very small fraction of its service life – less than a tenth of a percent. The relaying computer can thus take over these other tasks at practically no extra cost.

With the programmability and communication capability, the computer based relay offers yet another possible advantage that is not easily realizable in a conventional system. This is the ability to change relay characteristics (settings) as system conditions warrant it. More will be said about this aspect (adaptive relaying) in Chapter 10.

The high expectations for computer relaying have been mostly met in practical implementations. It is clear that most benefits of computer relaying follow from the ability of computers to communicate with various levels of a control hierarchy. The full flowering of computer relaying technology therefore has only been possible with the arrival of an extensive communication network that reaches into major substations. Preferably, the medium of communication would be fiber optic links with their superior immunity to interference, and the ability to handle high-speed high-volume data. It appears that the benefits of such a communication network would flow in many fields, and as more such links become available, the computer relays and their measurement capabilities become valuable in their own right. Where extensive communication networks are not available, many of the expected benefits of computer relaying must remain unrealized.

Other issues which are specific to computer relaying technology should also be mentioned. It has been noted that digital computer technology has advanced at a very rapid pace over the last twenty years. This implies that computer hardware has a relatively short lifespan. The hardware changes significantly every few years, and the question of maintainability of old hardware becomes crucial. The existing relays have performed well for long periods – some as long as 30 years or more. Such relays have been maintained over this period. It is difficult to envision a similar lifespan for computer based equipment. Perhaps a solution lies in the modularity of computer hardware; computers and peripherals belonging to a single family may provide a longer service life with replacements of a few modules every few years. As long as this can be accomplished without extensive changes to the relaying system, this may be an acceptable compromise for long service life. However, the implications of rapidly changing computer hardware systems are evident to manufacturers and users of this technology.

Software presents problems of its own. Computer programs for relaying applications (or critical parts of them) are usually written in lower level languages, such as assembly language. The reason for this is the need to utilize the available time after the occurrence of a fault as efficiently as possible. Relaying programs tend to be computation and input-output bound. The higher level languages tend to be inefficient for time-sensitive applications. It is possible that in time, with computer instruction times becoming faster, the higher level languages could replace much of the assembly language programming in relaying computers. The problem with machine level languages is that they are not transportable between computers of different types. Some transportability between different computer models of the same family may exist, but even here it is generally desirable to develop new software in order to take advantage of differing capabilities among the different models. Since software costs are a very significant part of computer relaying development, non-transferability of software is a significant problem.

In the early period of computer relaying development, there was some concern about the harsh environment of electric utility substations, in which the relays must function. Extremes of temperature, humidity, pollution as well as very severe EMI must be anticipated.

Another concern often raised by users of computer relays can be traced to the wide range of problems these relays can handle. It is rare to find a computer relay which does not require very large number of settings before it can be installed and commissioned. Where the organization using these devices has ample staff dedicated to working with computer relays, handling the complexity of setting these relays is not a problem. However, where the organization is small and a specialized staff for these applications cannot be justified, setting of these relays correctly and maintaining them for future modifications becomes a difficult task. Furthermore, if relays of different manufacture are in use within a single organization, it may become necessary to have experts who can deal with devices of different manufacture. Several Working Groups and Technical Committees of the Power Engineering Society of IEEE have attempted to develop a common user-interface to relays of different manufacture, but this task seems to be too complex and not much progress has been made in this direction.

1.4 Computer relay architecture

Computer relays consist of subsystems with well defined functions. Although a specific relay may be different in some of its details, these subsystems are most likely to be incorporated in its design in some form. Relay subsystems and their functions will be described next.

The block diagram in Figure 1.1 shows the principal subsystems of a computer relay. The processor is central to its organization. It is responsible for the execution of relay programs, maintenance of various timing functions, and communicating with its peripheral equipment. Several types of memories are shown in

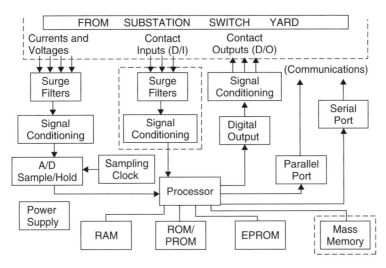

Figure 1.1 Subsystems of a relaying computer. The dashed line at the top indicates the boundary of the out-door switch yard. All other equipment is inside the control house

Figure 1.1 – each of them serves a specific need. The Random Access Memory (RAM) holds the input sample data as they are brought in and processed. It may also be used to buffer data for later storage in a more permanent medium. In addition, RAM is needed as a scratch pad to be used during relay algorithm execution. The Read Only Memory (ROM) or Programmable Read Only Memory (PROM) is used to store the programs permanently. In some cases the programs may execute directly from the ROM, if its read time is short enough. If this is not the case, the programs must be copied from the ROM into the RAM during an initialization stage, and then the real-time execution would take place from the RAM. The Erasable PROM (EPROM) is needed for storing certain parameters (such as the relay settings) which may be changed from time to time, but once set must remain fixed, even if the power supply to the computer is interrupted. Either a core type memory or an on-board battery backed RAM may be suitable for this function.

A large capacity EPROM is likely to become a desirable feature of a computer relay. Such a memory would be useful as an archival data storage medium, for storing fault related data tables, time-tagged event logs, and audit trails of interrogations and setting changes made in the relay. The main consideration here is the cost of such a memory. The memory costs have dropped sufficiently by now so that archival storage of oscillography and sequence-of-event data on a large scale within the relays has become possible.

Consider the analog input system next. At the outset it should be pointed out that Figure 1.1 is based upon using conventional transducers. If electronic CTs and CVTs are used, the input circuits may be significantly different and data are likely to be entered directly in the processor memory. The relay inputs are currents and voltages and digital signals indicating contact status. The analog signals must be converted to voltage signals suitable for conversion to digital form. This is done by the Analog to Digital Converter (ADC). Usually the input to an ADC is restricted to a full scale value of ± 10 volts. The current and voltage signals obtained from current and voltage transformer secondary windings must be scaled accordingly. The largest possible signal levels must be anticipated, and the relation between the rms (root mean square) value of the signal and its peak must be reckoned with. It is not necessary to allow for high frequency transients in most cases, as these are removed by anti-aliasing filters which have a low cut-off frequency. An exception to this is a wave relay, which does use the high frequency (traveling wave) components. For such relays (to be discussed more fully in Chapter 9), the scaling of signals must be such that the entire input signal with its largest anticipated high frequency component must not exceed the ADC input range.

The current inputs must be converted to voltages – for example by resistive shunts. As the normal current transformer secondary currents may be as high as hundreds of amperes, shunts of resistance of a few milliohms are needed to produce the desired voltage for the ADCs. An alternative arrangement would be to use an auxiliary current transformer to reduce the current to a lower level. However, any inaccuracies in the auxiliary current transformer would contribute to the total error of the conversion process, and must be kept as low as possible. An auxiliary current transformer serves another function: that of providing electrical isolation between the main CT secondary and the computer input system. In this case, the shunt may be grounded at its midpoint in order to provide a balanced input to following amplifier and filter stages. These considerations are illustrated in Figure 1.2(a) and (b).

Figure 1.2(c) shows connections to the voltage transformer. A fused circuit is provided for each instrument or relay, and a similar circuit may be provided for the computer relay as well. The normal voltage at the secondary of a voltage transformer is 67 volts rms for a phase to neutral connection. It can be reduced to the desired level by a resistive potential divider sized to provide adequate source impedance to drive the following stages of filters and amplifiers.

Although an auxiliary voltage transformer may be used in this case to provide additional isolation, it is not a necessity. Digital inputs to the computer relay are usually contact status, obtained from other relays or subsystems from within the substation. If the other subsystems are computer based, then these signals can be input to the computer relay without any special processing. An exception to this may be an opto-isolation circuit provided to maintain isolation between the two systems. When the digital inputs are derived from contacts within the yard (or control house), it is necessary to apply surge filtering and (or) optical isolation in order to isolate the computer relay from the harsh substation environment. Surge suppression for analog and digital signals is discussed next.

Suppression of surges from wiring connected to any protection system is a specialized subject with considerable literature of its own.[12,13] High voltage and high energy content surges are coupled into the wiring which connects current, voltage, and digital inputs to the protection system. The surges are created by faults

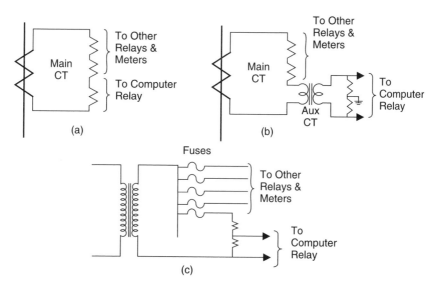

Figure 1.2 Scaling of current and voltage signals for input to the relay. (a) Direct connection in the main CT secondary. (b) Use of auxiliary CT. (c) Voltage transformer and potential divider

and switching operations on the power system, or by certain types of switching operations within the control house. For example, sparking contacts in inductive protection and control circuits within the control house have been found to be a source of very significant disturbances.[14] (See Chapter 7 for additional details). Suppression of these surges requires very careful grounding and shielding of leads and equipment, as well as low-pass filtering. Nonlinear energy absorbing Metal Oxide Varistors (MOVs) may also be used. Surge suppression filters are necessary for input and output wiring, as well as for the power supply leads.[12]

The ADC and anti-aliasing filter associated with the sampling process will be considered in greater detail in Sections 1.5 and 1.6. At this stage it is sufficient to be aware of their function in the overall relaying process. The anti-aliasing filters are low-pass analog filters designed to suit a specific choice of sampling rate used. The sampling instants are determined by the sampling clock, which must produce pulses at a fixed rate. The relationship between the sampling clock and several of the measurement functions performed by a computer relay is discussed in Chapters 9 and 10. For the present, it is sufficient to understand that, at each instant defined by the clock, a conversion from the instantaneous value of an analog input signal (voltage or current) to a digital form is performed by the ADC, and made available to the processor. Since the relay in general requires several inputs, several conversions are performed at each sampling instant. It is desirable (although not essential) that all signal samples be simultaneous, which means that either the conversion and transmission to the processor of each sample be very fast, or all the signals be sampled and held at the same instant for processing by a relatively

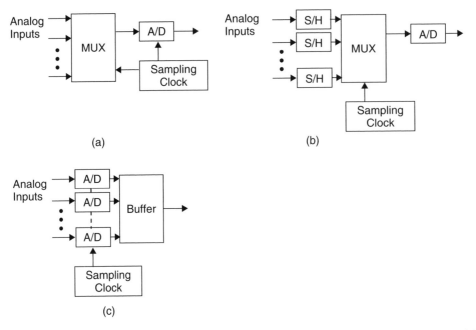

Figure 1.3 Multiple signal sampling process and its organization. (a) Single ADC with multiplexed input. (b) Sample and hold added to each channel. (c) Separate ADC for each channel

slow conversion-transmission cycle for each sample. This is typical of a multiplexed analog input system. A third option, technically feasible but expensive. is to use individual ADCs for each input channel. Trends in the ADC development and their reduced costs seem to point to the use of individual ADCs for each signal to be the preferred system. These options are illustrated in Figure 1.3.

It is well to consider this need for simultaneity in a little more detail at this point. Most relay functions require simultaneous measurement of two or more phasor quantities.

As will be seen in Chapters 3 and 8, the reference for these phasors is determined by the instant at which a sample is obtained. Thus if the phasors for signals x(t) and y(t) are computed from their samples beginning at instants t_x and t_y, the references for the two phasors will differ from each other by an angle θ, where

$$\theta = (t_x - t_y)\frac{2\pi}{T} \quad \text{radians}$$

where T is the fundamental frequency period of the signal. If the difference between t_x and t_y is known, then the phase angle between the two references is also known, and the two phasors could always be put on a common reference by compensating for θ. It would thus appear that simultaneous sampling of various input signals is not necessary, as long as the difference between the two is known and compensated for. On the other hand, all computations become much simpler if θ is zero and no compensation is needed. Furthermore, when needed, the samples of different signals could be combined directly (as in the case of a differential relaying application, where all input current samples of different signals could be added directly to form samples of the differential current). To be able to combine the samples directly, it is essential that the samples be taken simultaneously – and this fact, plus the relative ease of achieving it, has led to the general practice of simultaneous sampling of all input signals by each relaying computer. Indeed, there are benefits to be gained by coherently sampling all the quantities within a station as well as at all the stations within the system. System-wide synchronization will be considered in Chapters 8, 9 and 10.

Consider the sampling scheme shown in Figure 1.3(a). In the absence of sample-and-hold circuits, the different signal samples are obtained sequentially, and are not truly simultaneous. One period of a 60 Hz wave is 16.67 milliseconds. This corresponds to about 21.6 degrees per millisecond. Thus if the entire sampling scan can be completed in about 10 microseconds, the worst error created by sequential sampling amounts to about 0.2 degree – a negligible amount of error in any relaying application. Indeed, total scan periods of about 50 microseconds could be tolerated. In fact, a tolerance of 10–50 microseconds provides a good measure for describing any data samples as being simultaneous.

It should be mentioned that if simultaneous sampling is not possible, and yet it is needed for a relaying application, one could generate approximate simultaneous samples from non-simultaneous samples. Suppose that samples $x_k = \{x_1, x_2, \ldots x_n\}$ are obtained at instants $t_k = \{t_1, t_2, \ldots t_n\}$, whereas samples at $t'_k = t_k + \Delta T$ are needed. If $x(t)$ is assumed to be suitably band-limited, x'_k can be generated by interpolation formulas. The simplest procedure would be to use linear interpolation:

$$x'_k = x_k + (x_{k+1} - x_k)\frac{\Delta T}{t_{k+1} - t_k}$$

where $k = 1, 2, \ldots n - 1$ as shown in Figure 1.4. Higher order polynomials or spline functions may be used to obtain x'_k from x_k. Details may be found in any textbook on numerical methods.[15] It should be remembered that, in the context of relaying applications, any but the simplest linear interpolation formula would require excessively long real-time computation.

Returning once again to Figure 1.1, digital output from the processor is used to provide relay output in the form of open or close contacts. A parallel output port of the processor provides one word (typically two bytes) for these outputs. Each bit can be used as a source for one contact. The computer output bit is a Transistor to Transistor Logic (TTL) level signal, and would be optically isolated before driving a high speed multi-contact relay, or thyristors, which in turn can be used to activate external devices such as alarms, breaker trip coils, carrier control etc.

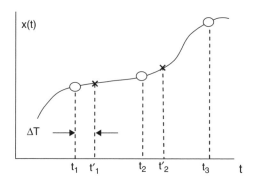

Figure 1.4 Interpolation for obtaining synchronous samples. Samples shown with x's are to be determined from samples actually obtained, which are shown by o's

Finally, the power supply is usually a single DC input multiple DC output converter powered by the station battery. The input is generally 125 volts DC, and the output could be 5 volts DC and ±15 volts DC. Typically the 5 volt supply is needed to power the logic circuits, while the 15 volt supply is needed for the analog circuits. The station battery is of course continuously charged from the station AC service.

1.5 Analog to digital converters

The Analog to Digital Converter (ADC) converts an analog voltage level to its digital representation. The principal feature of an ADC is its word length expressed in bits. Ultimately this affects the ability of the ADC to represent the analog signal with a sufficiently detailed digital representation. Consider an ADC with 12 bit word length – which, along with the 16 bit converter – is the most common word length in commercially available ADCs of today. Using a two's complement notation, the binary number 0111 1111 1111 (7FF in hexadecimal notation) represents the largest positive number that can be represented by a 12 bit ADC, while 1000 0000 0000 (800 in hexadecimal notation) represents the smallest (negative) number. In decimal notation, hexadecimal 7FF is equal to $(2^{11} - 1) = 2047$, and hexadecimal 800 is equal to $-2^{11} = -2048$. Considering that the analog input signal may range between ±10 volts, it is clear that each bit of the 12 bit ADC word represents 10/2048 volts, or 4.883 millivolts. Table 1.1 shows input voltages and their corresponding converted values in two's complement and decimal equivalent for 12 and 16 bit ADCs.

The equivalent input change for one digit change in the output (4.883 millivolts in case of a 12 bit A/D converter) is an important parameter of the ADC. It describes the uncertainty in the input signal for a given digital output. Thus an output of hexadecimal 001 represents any input voltage between 2.442 and 7.352 millivolts. This is the quantization error of the ADC. In general, if the word length of the ADC

Table 1.1 Two's complement 12 bit and 16 bit ADC input-outputs. maximum input voltage is assumed to be 10 volts

Input Volts	12 bit ADC		16 bit ADC	
	Hexadecimal	Decimal	Hexadecimal	Decimal
9.995	7FF	2047	7FFF	32767
5.0	400	1024	4000	16384
3.0	266	614	2666	9830
2.0	198	408	1999	6553
0.0	000	0	0000	0
−1.0	F33	−205	FCCB	−3277
−5.0	C00	−1024	BFFF	−16384
−10.0	800	−2048	8000	−32768

is N bits, and the maximum input voltage for the ADC is V volts, the quantization error q is given by

$$q = \frac{V}{2 \times 2^{N-1}} = 2^{-N} V$$

and normalized to the largest possible input voltage of V, the per unit quantization error is

$$\text{per unit} \quad q = 2^{-N}$$

clearly, the larger the number of bits in a converter word, the smaller is the quantization error.

Besides the quantization error, the ADC is prone to other errors as well. In order to understand the source of these errors, it is helpful to examine the principle of operation of an ADC.

1.5.1 Successive approximation ADC

A common type of analog-to-digital converter is the successive approximation ADC. Detailed information about this type of ADC and its design can be found in the literature.[16] The analog signal is amplified through an adjustable gain amplifier, as shown in Figure 1.5. A Digital to Analog Converter (DAC) converts the digital number in the output register of the ADC to an analog value. This signal is compared with the input analog signal, and the difference is used to drive up the count in the ADC output register. When the output of the DAC is within the quantization range of the analog input, the output is stable, and is the converted value of the analog signal. The amplifier is a source of additional error in the ADC. It may have a DC offset error as well as a gain error. In addition, the gain may have nonlinearity as well.

The combined effect of all ADC errors is illustrated in Figure 1.6. The offset error produces a shift in the input-output characteristic, whereas the gain error

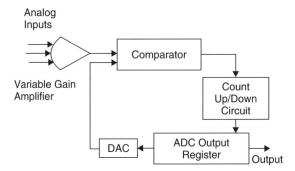

Figure 1.5   Successive approximation ADC. When the output of the comparator is positive, the ADC output register is incremented; when negative, it is decremented. When the
comparator output is less than the quantization error, the output is declared valid

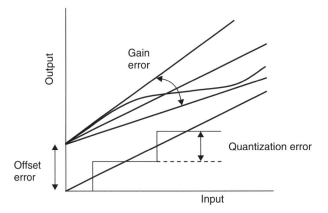

Figure 1.6   The effect of gain error, gain nonlinearity, and quantization on total error of
the ADC

produces a change in the slope. The nonlinearity produces a band of uncertainty in
the input-output relationship. If the gain error and the nonlinearity error are bounded
by two straight lines as shown in Figure 1.6, the total error in the ADC for a given
voltage input V is given by ε_v:

$$\varepsilon_v = K_1 \times FS + K_2 \times V$$

where FS is the full scale value of the input voltage, and K_1 and K_2 are constants
depending upon the actual uncertainties of the conversion process.

   It is possible for the gain setting to be changed between samples, although at sampling rates corresponding to relaying applications (of the order of 1 kHz), dynamic
gain changing would be too time consuming. Consequently, the error model given
above is fairly representative of the ADCs used in relaying applications. It should

also be clear that, when the input signal is a small fraction of the full scale value, the first error term is dominant and the error at every sample is likely to be of the same size. On the other hand, when the input signals approach the full scale, the second term may dominate and each sample error may be proportional to its nominal value.

The quantization error component of the total error is a random process, while the remaining error components are deterministic errors depending upon the gain, offset and non-linearity errors present at a given moment. If we consider the ensemble of all operating conditions under which the relay must operate, this too may be treated as a random process. The error model given for ε_v above should then be understood to represent the standard deviation σ_v of the random measurement noise. Such error models will be considered in Chapter 3.

1.5.2 Delta-sigma ADC

The Delta-Sigma analog-to-digital converter has become the ADC of choice in recent years.[17] These converters use a 1 bit analog to digital converter, thus making the analog signal processing simple and inexpensive. A very high sampling rate is used (over-sampling) and the digital signal processing is used to provide appropriate anti-aliasing filters and decimation filters. The digital circuitry in these ADCs are more complex, but are relatively inexpensive to manufacture.

The block diagram of a generic delta-sigma ADC is shown in Figure 1.7(a). The signal x_1 is obtained by subtracting from the input signal x the output of the 1 bit ADC (y) converted to analog form by the 1 bit Digital-to-Analog Converter. The

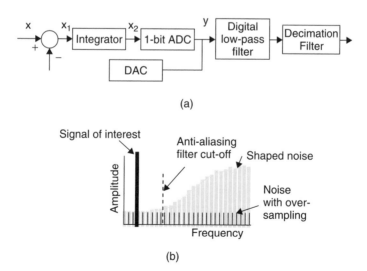

(a)

(b)

Figure 1.7 (a) Block diagram of a delta-sigma analog-to-digital converter. (b) Noise reduction in ADC output by shaping the filter characteristic

signal x_1 is integrated to produce the signal x_2 which is fed to the 1 bit ADC. The feed-back circuit ensures that the average of the analog input is equal to that of the converted signal. The output of the 1 bit ADC is a 1 bit data stream clocked at high frequency (over-sampling). A digital low-pass filter converts the 1 bit data stream to a multi-bit data stream, which is finally filtered by a decimation filter to achieve the sampling rate of interest in relaying applications. For example, with an over-sampling frequency of 40 kHz, a decimation filter output at 2 kHz could be obtained with a 16 bit resolution. The over-sampling reduces the amount of noise in the frequency band of interest, as it spreads the signal noise equally throughout the bandwidth corresponding to the over-sampling rate. A further reduction in noise in the output is achieved by shaping the digital filter characteristic so that the noise is concentrated at the high end of the spectrum, which in turn is eliminated by the anti-aliasing filter (Figure 1.7(b)).

1.6   Anti-aliasing filters

The need for anti-aliasing filters will be established in Chapter 3. For the present, we will accept that these are low-pass filters with a cut-off frequency equal to one-half the sampling rate used by the ADC. An ideal anti-aliasing filter characteristic with a cut-off frequency f_c is shown in Figure 1.8. A practical filter can only approximate this 'brick-wall' shape, as shown by the dotted line in Figure 1.8. Next, we will consider design aspects of practical anti-aliasing filters.

Anti-aliasing filters could be passive, consisting of resistors and capacitors exclusively; or active, utilizing operational amplifiers. As some buffering between the filters and the ADC is generally necessary, an operational amplifier is needed in any case, and one could use the active filter design which leads to smaller component sizes. An active filter may also be designed using the monolithic hybrid microelectronic technology providing compact packaging. The transfer function for the filter in any case is determined from considerations of sharpness of cut-off in the stop band, and the transient response of the filter.

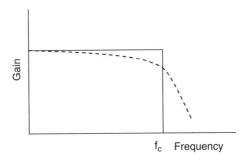

Figure 1.8   Ideal anti-aliasing filter characteristic for a cut-off frequency of f_c. Approximate realizable characteristic shown by the dotted line

In general, if filters with very sharp cut-off are employed, they produce longer time delays in their step function response.[18] In most applications of computer relaying, two-stage RC filters are found to provide an acceptable compromise between sharpness of the cut-off characteristic in the stop band, and the time delay in their step input response. A second order Butterworth, Chebyshev, or maximally flat (Bessel) filter may be used to satisfy computer relaying requirements. However, these filters have a significant overshoot in their step input response. As an example, we will consider the design of a two-stage RC filter suitable for a sampling process using a sampling rate of 720 Hz (12 times the fundamental frequency for a 60 Hz power system). The filter must have a cut-off frequency of 360 Hz. We may further specify a DC gain of unity – which makes either an active or a passive design possible. An active filter can of course be designed to provide any other reasonable gain.

Two-stage RC filters are quite popular because of their simplicity, passive components, and a reasonable frequency response. They suffer from the disadvantage that they produce a rounded characteristic at the beginning of the stop band. A two-stage RC filter achieves a 12 db per octave attenuation rate when it is well into its stop band. Indeed, this is a property of an all-pole second order filter.[18] The transfer function of a two-stage RC filter is given by:

$$H(j\omega) = \frac{1}{1 + j\omega(R_1C_1 + R_2C_2 + R_1C_2) - \omega^2(R_1C_1R_2C_2)}$$

R_1, C_1, R_2, C_2 being the components of two stages. These components must be adjusted to provide the necessary attenuation at a desired cut-off frequency f_c. Figure 1.9(a) shows a two-stage RC circuit with this transfer function and a cut-off frequency of 360 Hz. The frequency response and step wave response of this filter are shown in Figures 1.9(b) and (c). As can be seen, the step wave response is reasonable, producing an essentially correct output in about 0.8 millisecond after application of the step wave. The phase lag at the fundamental power frequency (60 Hz) is about 11 degrees, which corresponds to a time delay of about 0.7 millisecond. Considering that this filter has been designed for a sampling frequency of 720 Hz, the phase delay produced by it is about one-half the sampling period. Recall that the sampling period at a sampling frequency of 720 Hz is 1.388 milliseconds.

Second order Chebyshev filters produce a somewhat steeper initial cut-off in their stop band. However, this is at the expense of a ripple in the pass band. The step wave response of Butterworth and Chebyshev filters is somewhat poorer, having a significant overshoot. A comparison of the frequency response and step wave response of these three second order filters with a cut-off frequency of 360 Hz is shown in Figure 1.10. Figure 1.11 shows an active realization of the two-pole Butterworth filter of Figure 1.9. Note that there are no inductors in this realization whereas a passive Butterworth filter must use inductances. This may well be one of the considerations in the final choice of active or passive filter design.

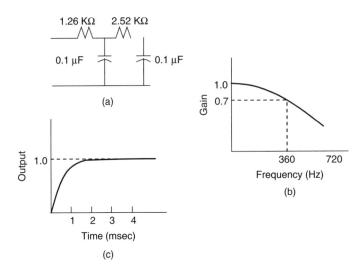

Figure 1.9 Two-stage RC filter with a cut-off frequency of 360 Hz. (a) RC ladder realization. (b) Frequency response. (c) Step wave input response

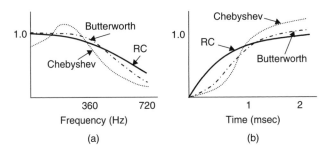

Figure 1.10 Comparison of second order RC, Butterworth, and Chebyshev filters with a cut-off frequency of 360 Hz. (a) Frequency response. (b) Step wave response

Another consideration in selecting a filter design is the stability of its transfer function in the presence of variation in component values due to aging and temperature variations. Consider the passive two-stage RC filter of Figure 1.9. Its gain and phase shift are obtained by taking the magnitude and phase angle of $H(\omega)$:

$$G(\omega) \equiv |H(\omega)|$$

$$= \sqrt{\frac{1}{\{1 - \omega^2(R_1C_1R_2C_2)\}^2 + \omega^2(R_1C_1 + R_2C_2 + R_1C_2)\}^2}}$$

$$\phi(\omega) \equiv \angle H(\omega)$$

$$= -\arctan\left[\frac{\omega(R_1C_1 + R_2C_2 + R_1C_2)}{1 - \omega^2(R_1C_1R_2C_2)}\right]$$

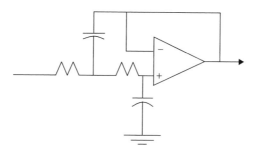

Figure 1.11 Active circuit realization of a two-pole Butterworth filter with a cut-off frequency of 360 Hz

If we consider that all four components used in the passive circuit may vary by small amounts, the variations in the gain and phase shift are given by

$$\Delta G(\omega) = \frac{\partial G}{\partial R_1}\Delta R_1 + \frac{\partial G}{\partial R_2}\Delta R_2 + \frac{\partial G}{\partial C_1}\Delta C_1 + \frac{\partial G}{\partial C_2}\Delta C_2$$

$$\Delta \phi(\omega) = \frac{\partial \phi}{\partial R_1}\Delta R_1 + \frac{\partial \phi}{\partial R_2}\Delta R_2 + \frac{\partial \phi}{\partial C_1}\Delta C_1 + \frac{\partial \phi}{\partial C_2}\Delta C_2$$

Taking the partial derivatives at the selected nominal values of R_1, R_2, C_1 and C_2 in Figure 1.9, we get

$$\frac{\Delta G}{G} = -0.013\frac{\Delta R_1}{R_1} - 0.013\frac{\Delta R_2}{R_2} - 0.004\frac{\Delta C_1}{C_1} - 0.022\frac{\Delta C_2}{C_2}$$

$$\frac{\Delta \phi}{\phi} = 0.493\frac{\Delta R_1}{R_1} + 0.493\frac{\Delta R_2}{R_2} + 0.242\frac{\Delta C_1}{C_1} + 0.734\frac{\Delta C_2}{C_2}$$

Often it is the relative deviation in the phase angle $\Delta\phi/\phi$ at 60 Hz due to changes in resistor and capacitor values which is of greatest concern, since this is greater in magnitude than the relative deviation in gain $\Delta G/G$. As can be seen from Problem 1.3, the gain magnitude and phase angle of an active filter are more sensitive to variations in component values as compared to those of a passive filter. Component variations can be kept small by selecting high precision metal film resistors and polystyrene or polycarbonate capacitors.

1.7 Substation computer hierarchy

Let us consider the hierarchy of various relaying and other computers in a substation. Computer relays are expected to be a part of a system wide protection and control computer hierarchy.[19,20] Functionally the hierarchy structures that are being planned for implementation may be represented as in Figure 1.12. Relay computers and their input-output systems are at the lowest level of this hierarchy, and

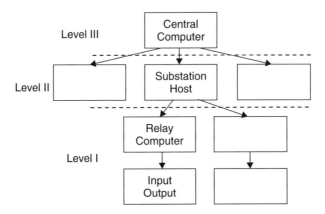

Figure 1.12 System wide hierarchical computer system. Computer relays are at the lowest level of the hierarchy. They are linked with substation host computer which in turn is linked with a system center computer

communicate with the switch yard through the relay input and output signals. As the relay outputs are connected to circuit breakers and could also be connected to remote controlled switches in the yard, the relay may serve as a conduit for supervisory control tasks at the substation. The control commands flow from the system center, through the substation host computer and to the relay computers. All the relay computers within the station are linked to the substation host computer. This host acts as a data concentrator for all historical and oscillography records collected by the relay computers. It – along with all other substation computers – transmits these data to the system central computer. The substation host computer also provides an interface between the relay computers and the station operators. Through this interface the relay settings, calibration, target interrogation, or diagnostic and maintenance functions can be performed. The substation host computer may also be used to produce some coordinated sequence-of-events for the substation as an aid to station maintenance personnel.

The role of the central computer is even less critical in the conventional relaying process. It initiates various supervisory control commands at the behest of the operator. It also collects historical data from all the substation computers and creates oscillography, coordinated sequence of event analyses, and various book-keeping functions regarding the operations performed at the station. The central computer will play a more direct role if adaptive relaying becomes accepted by the relaying community. Adaptive relaying principles and their relationship to computer relaying have been discussed in some recent publications.[21–23]

The functions of various computers in the system wide computer hierarchy may be summarized as follows:

Level I: Relaying, input output to the switch yard, measurements, control, diagnostics, man-machine interface, communications with level II.

Level II: Man-machine interface, data acquisition and storage, sequence of events analyses and coordination, assignment for back-up in case of failures, communication with level I and level III computers.

Level III: Initiate control actions, collect and collate system wide sequence of event analyses, communication with level II, oscillography and report preparation, adaptive relaying.

1.8 Summary

In this chapter, we have traced the development of computer relaying. We have examined the motivation behind this development, and the expectations for systems of the future. Computer based relays have become the standard of the electric power industry. Their service record in field installations has been comparable to that of traditional relays, and in many respects they offer advantages which are not readily available with traditional relays. Examples of such features are adaptive relaying (to be considered in Chapter 10), oscillography records saved for post-mortem analyses, and the ability to address the relays from remote sites through available communication links. In fact, the communication capability is the single most valuable asset of computer based relays. Of course, the communication capability brings with it the concern for security from malicious intervention from unauthorized persons. As with most computer based systems, this concern can be alleviated by appropriate fire-walls and other security measures.

In this chapter we have considered in detail the functional block diagram of a computer relay and its place in the hierarchical system wide computer system. We have discussed the Analog to Digital conversion process, and the sources of errors in data conversion. We have also discussed the anti-aliasing filter design, and its contribution to the overall error in the relay input system. The Problems that follow further illustrate many of these concepts.

Problems

1.1 A 500 kV transmission line has a normal load current of 1000 amperes primary, and a maximum symmetrical fault current of 30 000 amperes. Determine the CT and CVT ratios and ohmic values of shunts and potential dividers in their respective secondary circuits. Assume that a full DC offset may occur in the fault current, and allow for a dynamic overvoltage of 20%. Determine the smallest load current that the input system can read if the ADC used is a 12 bit converter.

1.2 Consider the non-simultaneous sampling of two input signals. Assume the difference between sampling instants for the two signals to be 30% of the sampling interval. What is the error in the sample of one signal calculated to be simultaneous with the sample of the other signal using a linear interpolation formula? Assume that both input signals are of pure fundamental frequency.

Assume suitable sampling rates, and arbitrary placement of sampling instants on the waveforms.

1.3 A phasor is calculated from samples of an input signal taken over one period. If the samples are obtained with a 12 bit ADC, what is the error in the phasor computation due to the quantization error? Obtain the greatest upper bound for the error in magnitude and phase angle of the phasor. The phasor calculation is explained in Chapter 3. For the present, assume that the phasor is the same as the fundamental frequency component calculated by the familiar Fourier series formula.

1.4 Repeat Problem 1.3 for the errors introduced by an imperfect input amplifier having an offset error, a gain error, and gain nonlinearity. Assume suitable bounds for these errors.

1.5 Design a third order Butterworth filter with a cut-off frequency of 360 Hz. Plot its transient response to a step input function. Also plot its frequency response for a frequency range of DC to 720 Hz.

1.6 Design a two-stage RC filter with a cut-off frequency of 300 Hz, and compare its frequency and step input response with those of the two-stage RC filter having a cut-off frequency of 360 Hz. Often a 300 Hz cut-off frequency filter may be used for anti-aliasing purposes where sampling frequency is 720 Hz. This gives better noise immunity to the filter, but it is of course not suitable when it is necessary to calculate the fifth harmonic of the input signal – for example, in a transformer relay.

1.7 Complete the design of the second order Butterworth filter by finding the values of the components in Figure 1.10.

1.8 Verify the expressions for the relative gain and phase angle variations $\Delta G/G$ and $\Delta\phi/\phi$ given in Section 1.6 for the two-stage RC filter. Determine a general formula, and then substitute the values of the resistors and capacitors from Figure 1.8(a).

1.9 Derive expressions similar to those in Problem 1.8 for the two-stage Butterworth filter, and obtain numerical results for the filter design of Problem 1.5 Verify that, for the same relative variation in component values, the Butterworth filter has greater variation in its gain and phase shift when compared to the two-stage RC filter.

References

[1] Rockefeller, G. D. (1969) Fault protection with a digital computer, IEEE Transactions on Power Apparatus and Systems (IEEE Trans. on PAS), vol. 88, no. 4, pp. 438–461.
[2] Mann B. J. and Morrison, I. F. (1971) Digital calculation of impedance for transmission line protection, IEEE Trans. on PAS, vol. 90, no. 1, pp. 270–279.

[3] Poncelet, R. (1972) The use of digital computers for network protection, CIGRÉ Paper no. 32-08.

[4] Takagi, T., Baba, J., Uemura K., and Sakaguchi, T. (1977) Fault protection based on traveling wave theory – Part I: Theory, IEEE Summer Power Meeting, Mexico City, paper no. A77 750-3.

[5] Dommel, H. W. and Michels, J. M. (1978) High speed relaying using travelling wave transient analysis, IEEE PES Winter Power Meeting, New York, January 1978, Paper No. A78 214-9.

[6] Cory B. J. and Moont, J. F. (1970) Application of digital computers to busbar protection, IEEE Conference on the Application of Computers to Power System Protection and Metering, Bournemouth, England, May 1970, pp. 201–209.

[7] Sykes, J. A. and Morrison, I. F. (1972) A proposed method of harmonic restraint differential protection of transformers by digital computers, IEEE Trans. on PAS, vol. 91, no. 3, pp. 1266–1272.

[8] Sachdev, M. S. and Wind, D. W. (1973) Generator differential protection using a hybrid computer, IEEE Trans. on PAS, vol. 92, no. 6, pp. 2063–2072.

[9] Phadke, A. G., Horowitz, S. H., Thorp, J. S. (1983) Integrated computer system for protection and control of high voltage substations, CIGRÉ Colloquium, Tokyo, Japan, November 1983.

[10] Deliyannides J. S. and Udren, E. A. (1985) From concepts to reality: the implementation of an integrated protection and control system, *Developments in Power System Protection*, IEE Conference Publication no. 249, London, April 1985, pp. 24–28.

[11] North American Reliability Council, System Disturbance, 1983, 1984 etc. Research Park, Terhune Road, Princeton, New Jersey.

[12] Surge Withstand Capability (SWC) Tests for Protective Relays and Relay Systems, P472/D9, C37.90.1-198x. Draft Document of the Power System Relaying Committee, June 8, 1987.

[13] Surge Withstand Capability (SWC) Tests, ANSI C37.90a, 1974.

[14] Kotheimer W. C. and Mankoff, L. L. (1977) Electromagnetic interference and solid state protective relays, IEEE Trans. on PAS, vol. PAS-96, no. 4, pp. 1311–1317.

[15] Henrici, P. (1964) *Elements of Numerical Analysis*, John Wiley & Sons, Ltd.

[16] *Product Data Book*, Burr-Brown Research Corporation, 1982, Tucson, Arizona.

[17] Candy, J. C. and Temes, G. C. (1991) Oversampling methods for data conversion, IEEE Pacific Rim Conference on Communications, Computers and Signal Processing, May 9–10, 1991.

[18] Chen, Wai-Kai (1986) *Passive and Active Filters*, John Wiley & Sons, Ltd, Chichester.

[19] Udren, E. A. (1985) An integrated, microcomputer based system for relaying and control of substations: design features and testing program, 12th Annual Western Protective Relaying Conference, Spokane, Washington, October 1985.

[20] Phadke, A. G. (Convener), CIGRÉ Working Group 34.02 Final Report, Paris, 1985.

[21] Thorp, J. S., Phadke, A. G., Horowitz, S. H. and Begovic, M. M. (1987) Some applications of phasor measurements to adaptive protection, Proceedings of PICA, Montreal.

[22] Horowitz, S. H., Phadke, A. G., Thorp, J. S. 1987 Adaptive transmission system relaying, IEEE PES Summer Power Meeting, San Francisco, 1987.

[23] Phadke, A. G., Thorp, J. S., Horowitz, S. H. (1987) Impact of adaptive protection on power system control, Proceedings of PSCC, Lisbon.

2

Relaying practices

2.1 Introduction to protection systems

In this chapter we will summarize the operating principles of different types of relays, and express their operating principles through their performance equations. The performance equation is a mathematical relationship between the input quantities of a relay and its output. Generally, the inputs are voltages and currents – sometimes supplemented by status of some contacts – and the outputs are status changes (on-off) of the output contacts of the relay. Conventional relay performance is described in terms of voltage and current phasor inputs, and we will follow the same practice in this chapter. However, it must be understood that the phasor concept implies steady-state fundamental frequency sinusoidal waveforms; while immediately following a fault the currents and voltages of a power system are rich in transient components of other than fundamental frequency. As will be seen in the next chapter, the phasor representation of an input quantity can be defined under these conditions, and this interpretation of the phasor under transient conditions will be understood in the performance equations derived in this chapter.

It should also be mentioned that the reason for reviewing the performance equations of conventional relays is to provide a point of reference for computer relaying development. It will be found that in most instances the conventional relaying equations provide the ideal method of achieving a protection goal. It is sound practice to examine ways of implementing conventional relaying principles in a digital computer before going on to newer developments. In this fashion, it can be assured that the job of meeting relaying needs – done admirably by most conventional relays – is also done by the computer relays. Only then can we begin to explore development of newer relaying principles. Although we will describe the operating principles of various relays, it should be noted that their application to power systems is a specialized field, and by and large we will not get into

Computer Relaying for Power Systems 2e   by A. G. Phadke and J. S. Thorp
© 2009 John Wiley & Sons, Ltd

application related matters. For example, we will describe how an overcurrent relay functions, but we will not describe how a set of overcurrent relays can be applied for transmission line protection. This is in keeping with our objective of describing the computer relaying technology, i.e. how computer relays work. For the most part, their application to solve a specific protection problem is similar to that of conventional relays. For a complete treatment of traditional relaying and application practices, the reader is referred to several excellent text books and reference manuals on protective relaying.[1-5] In addition, the IEEE Press has published an excellent collection of important papers dealing with all aspects of relaying.[6]

Earliest relays were electromechanical devices consisting of plungers, balanced-beams and induction discs or cups. With the exception of the balanced-beam types, all other electromechanical relays are in use at the present time. Electromechanical relays are robust – both mechanically and from the point of view of electromagnetic interference (EMI). Although they can be very fast (quarter cycle operating time), usually they are slow – their speed of operation is measured in cycles or seconds. They also require a fairly high amount of energy to operate, thereby necessitating current and voltage transformers with relatively high volt-ampere capability.

In the late 1950s solid-state relays began to appear. These were designed with discrete electronic components such as diodes, transistors and operational amplifiers. Early solid-state relays were plagued by failures of components due to EMI, and by failures brought on by the high failure rate of the early solid state components. To a certain extent, some relay engineers still consider solid state relays to be less reliable than their electromechanical counterparts. However, for most users solid-state relays became an important element of modern protection system design. Modern solid state relays are relatively maintenance-free and offer a greater flexibility as far as protection applications are concerned. Their operating speed is high – of the order of one cycle or less. On many power systems, the protection system consisted of a combination of solid state relays and electromechanical relays – electromechanical relays being more common in simpler applications such as overcurrent relaying, while solid-state relays dominated in more complex applications such as pilot relaying or step-distance relaying. In recent years, computer relays have supplanted solid state and electromechanical relays as protection systems of choice.

2.2 Functions of a protection system

A protection system protects the power system from the deleterious effects of a sustained fault. A fault (meaning in most cases a short circuit, but more generally an abnormal system condition) occurs as a random event. If some faulted power system component (line, bus, transformer, etc.) is not isolated from the system quickly, it may lead to power system instability or break-up of the system through the action of other automatic protective devices. A protection system must

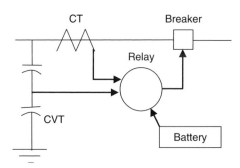

Figure 2.1 Subsystems of a protection system. Besides relays, the protection system consists of transducers, circuit breakers and station battery

therefore remove the faulted element from the rest of the power system as quickly as possible.

Although a protection system is usually understood to mean relays, it consists of many other subsystems which contribute to the fault removal process. These subsystems are identified in Figure 2.1. The circuit breaker actually isolates the faulted circuit by interrupting the current at or near current zero. A modern Extra High Voltage (EHV) circuit breaker can interrupt fault currents of the order of 100 000 amperes at system voltages of up to 800 kV. It can do this as quickly as the first current zero after the initiation of the fault, although more often it does so at the second or third current zero. The breaker is operated by energizing its trip coil from the station battery, and the relay(s) do this job by closing contacts between the battery and the trip-coil. Very often other relays (reclosing relays) are used to reclose the circuit breaker after a suitable time interval.

The transducers (current and voltage transformers, or CTs and CVTs) constitute another major component of the protection system. They are necessary because the high magnitude currents and voltages of the power system must be reduced to more manageable levels in order to drive low energy (and hence safe for human access) devices such as relays. We will consider the current and voltage transformers in some detail in Section 2.6. For the present, it is sufficient to note that certain features of the transducers have been standardized. Current transformer secondary rating has been standardized at 5 amperes or 1 ampere, the latter standard being more common in Europe. (A few other standard ratings also exist but they are not very common). This implies that the maximum load current in the primary winding of the current transformer would produce 5 amperes (1 ampere) or less in its secondary winding. This leads to a desired CT winding ratio, which is then approximated by one of the standard CT ratios available. The voltage transformers have the secondary windings rated at 67 volts phase-to-neutral. Within certain limits (as discussed in Section 2.6), the current and voltage transformers reproduce the primary current and voltage waveforms faithfully on their secondary side. The relay thus sees a scaled down version of currents and voltages that exist on the power system.

The last and most important component for our discussion of the protection system is the relay. This is a device which responds to the condition of its inputs (voltages, currents, or contact status) in such a manner that it provides appropriate output signals to trip circuit breakers when input conditions correspond to faults for which the relay is designed to operate. Relays are the logic elements in the entire protection system. The design of a relay (whether analog or digital) must be such that all fault conditions for which it is responsible produce a trip output, while no other conditions should. Much of this book will be dedicated to design techniques for relaying algorithms such that these requirements are met.

This is a good point to discuss the concept of **reliability** as understood in relaying literature. To a relay engineer, a reliable relay has two attributes: it is **dependable**, and it is **secure**. Dependability implies that the relay will always operate for conditions for which it is designed to operate. A relay is said to be secure if the relay will not operate for any other power system disturbance. Of the two attributes (dependability and security), the latter is more difficult to achieve. Every fault in the neighborhood of a relay will disturb its input voltages and currents. However, the relay should disregard those voltage and current conditions that are produced by faults which are not the responsibility of the relay.

The responsibility for protection of a portion of the power system is defined by a zone of protection. A zone of protection is a region clearly defined by an imaginary boundary line on the power system one line diagram. A protection system – consisting of one or several relays – is made responsible for all faults occurring within the zone of protection. When such a fault occurs, the protection system will activate trip coils of circuit breakers thereby isolating the faulty portion of the power system inside the zone boundary. Usually – though not always – the zones of protection are defined by circuit breakers. If the zone of protection does not have a circuit breaker at its boundary, the protection system must trip some remote circuit breakers (transfer the trip command through a communication channel) to de-energize the faulted zone. Figure 2.2 shows a portion of the power system divided into various zones of protection. Zones 1, 2, and 3 are transmission line protection zones for the various lines. A fault on any of these lines would be detected by their corresponding protection systems, and trip appropriate breakers at zone boundaries. Zone 4 is the bus protection zone. Zone 5 is the zone for transformer protection. Note that there is no circuit breaker at one end of this zone, and consequently the transformer protection system must trip the breaker at bus A, and through a communication channel remotely trip the breaker at bus C.

Note also that the zones of protection always overlap. This is in order to ensure that no portion of the system is left without primary high speed protection (i.e. there are no blind spots in the protection system). Although overlap is achieved in Figure 2.2 by including the circuit breaker in each neighboring zone, in reality this may not be possible under all circumstances. Zone overlap is achieved through the

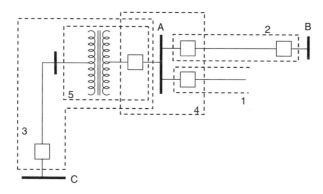

Figure 2.2 Zones of protection. Zone 2 defines the boundary for protection of transmission line A-B. Zone 4 defines bus-A protection. Zone 5 is a transformer protection zone

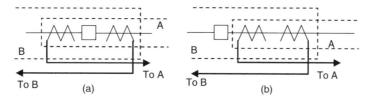

Figure 2.3 Principle of zone overlap. (a) When current transformers are available on either side of the breaker. (b) When a single current transformer with multiple secondary windings is available

proper choice of current transformers dedicated to each protection system. Consider the arrangement shown in Figure 2.3(a). A current transformer is assumed to exist on either side of the circuit breaker. In this case, the protection systems on either side of the breaker use current transformers from opposite sides of the circuit breaker.

When current transformers are not available on both sides of the circuit breakers, an overlap is achieved by using current transformer secondary windings on the far side as illustrated in Figure 2.3(b). In this case, although there is no blind spot in *relaying, tripping* for faults between the circuit breaker and the CT requires special consideration. It is desirable to keep the region of overlap as small as possible.

Occasionally, a zone will be protected by several protection systems in order to make sure that failure of the protection system itself will not leave the power system unprotected. This reinforces the dependability of the overall protection system. In such cases, it is desirable to have as much independence between the two protection systems as possible. It would be prohibitively expensive to duplicate the circuit breaker, the current transformer, the voltage transformer, or even the station battery. However, some degree of separateness may be obtained by using different secondary windings of a current transformer for the two protection systems, by using separate fuses in the voltage transformer circuit, by providing separate trip coils for the circuit breaker, and in exceptional cases by providing separate batteries for relays and breaker trip circuits. These attempts are made in order to avoid common failure modes among the different protection systems and thereby to improve the dependability of the entire protection system.

Almost all the relays in use on power systems may be classified as follows:

1. *Magnitude Relays*: These relays respond to the magnitude of the input quantity. An example is the overcurrent relay which responds to changes in the magnitude (either the peak value or the rms value) of the input current.
2. *Directional Relays*: These relays respond to the phase angle between two AC inputs. A commonly used directional relay may compare the phase angle of a current with a voltage. Or, the phase angle of one current may be compared with that of another current.
3. *Ratio Relays*: These relays respond to the ratio of two input signals expressed as phasors. Ratio of two phasors is a complex number, and a ratio relay may be designed to respond to the magnitude of this complex number or to the complex number itself. The most common ratio relays are the several versions of impedance or distance relays.
4. *Differential Relays*: These relays respond to the magnitude of the algebraic sum of two or more inputs. In their most common form, the relays respond to the algebraic sum of currents entering a zone of protection. This algebraic sum may be made to represent the current in any fault (if it exists) inside the zone of protection.
5. *Pilot Relays*: These relays utilize communicated information from remote locations as an input signal. This type of protection generally communicates the decision made by a local relay of one of the four types described above to relays at the remote terminals of a transmission line.

A functional description of input-output relationships of the five relay classes described above will give us sufficient basis for the design of relays – all else is application of these designs to a given protection problem. As far as digital relaying is concerned, we may then proceed directly to a discussion of relaying algorithms. However, we will give a very brief overview of the application of these relays to system protection problems. As mentioned in Section 2.1, the reader will do well to refer to several excellent books on application of relays to power systems. The material in Sections 2.4, 2.5, and 2.6 is offered in order to provide a degree of continuity in our development, as it is not advisable to develop a theory of relay algorithm design without some note being taken of the application principles.

2.3   Protection of transmission lines

2.3.1   Overcurrent relays

When a fault occurs on a power system, the fault current is almost always greater than the pre-fault load current in any power system element. A very simple and effective relaying principle is that of using the current magnitude as an indicator of a fault. Overcurrent relays (as such relays are known) can be used to protect practically any power system element, i.e. transmission lines, transformers, generators, or motors. We will use a transmission line as an example to illustrate the application. For a fault within the zone of protection, the fault current is smallest at the far end of the line and greatest at the relay (breaker) end. It is assumed that the system is radial – i.e. the source of power is only on the left side in Figure 2.4(a). If the minimum fault current possible within the zone of protection is greater than the maximum possible load current, it would be possible to define the operating principle of a relay as follows:

$$|I| \geq I_p \text{ fault in zone, trip}$$

$$< I_p \text{ no fault in zone, do not trip} \tag{2.1}$$

where I is the current in the relay and I_P is a setting described below. Note that the current magnitude must be derived from an AC waveform which may include a decaying DC component whose magnitude depends upon the instant of fault occurrence. Figure 2.4(b) is a sketch of the variation in steady state AC fault current – (known as the symmetrical fault current) with fault location. The relay characteristic given by Equation (2.1) is defined in terms of the symmetrical fault current.

The quantity I_P is known as the pickup setting of the relay. The above equation describes an ideal relay operating characteristic as shown in Figure 2.5(a). The relay does not operate (operating time is infinite) as long as the current magnitude is less than I_P. If the current magnitude exceeds I_P, the relay operates taking a time T_{min} to close its contacts. This type of relay is termed an instantaneous relay. It

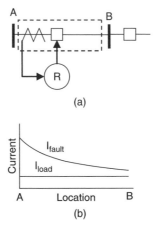

(a)

(b)

Figure 2.4 Overcurrent protection of transmission lines. (a) Radial system protection. (b) Fault current magnitude as a function of fault location

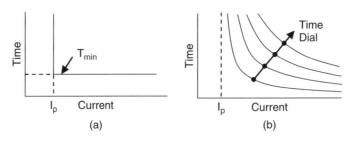

Figure 2.5 Overcurrent relay operating time. (a) Instantaneous relay, and (b) Time over-current relay

is often desirable to have the operating time depend upon the current magnitude: being smallest when the current is greatest as shown in Figure 2.5(b). Such a characteristic is known as an inverse-time characteristic and the relay is known as a time overcurrent relay. In practice, I_p would be selected to lie between maximum load current and minimum fault current.

A ratio of 3 between I_p and the minimum fault current is considered desirable for getting a well defined operating time for all faults. There is a great variety in the shape of the inverse-time characteristic: the shapes are described as being 'inverse', 'very inverse', and 'extremely inverse'. Admittedly, these are rather loose descriptions and the actual characteristic furnished by the manufacturer must be used in determining the relay setting. Furthermore, even for a relay of a given type, the operating time can be moved up (made slower) by turning the 'time dial' on the relay. This is also illustrated in Figure 2.5(b). By convention the lowest time dial setting (fastest operating time) is generally 1/2, and the slowest setting is 10. A typical time overcurrent relay characteristic of a commercially available relay is shown in Figure 2.6. The reader may want to work out some of the problems at the end of this chapter to gain insight into the application of time overcurrent relays and relay coordination.

2.3.2 Directional relays

When the power system is not radial (source on one side of the line as in Figure 2.4), an overcurrent relay may not be able to provide adequate protection. Consider the zone of protection for a line with sources behind both ends as shown in Figure 2.7(a). In such a case, depending upon the relative strength of the source on the two sides, it may be that for a fault such as F_1 (which is within the zone of protection of the transmission line), the current flowing through the relay at B is less than the current that would flow through the same relay (albeit in a reverse direction) for a fault at F_2 – which is outside its zone of protection. In such a case, an overcurrent relay set to trip for a fault at F_1 would also trip for a fault at F_2 – an unacceptable loss of security. This situation is resolved by providing a 'directional' relay at B (as well as at A) which does not operate when the fault is away from the zone of

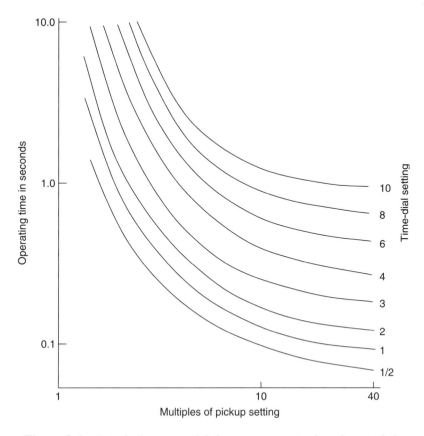

Figure 2.6   A typical commercial time overcurrent relay characteristic

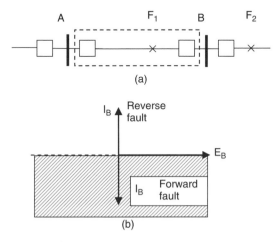

Figure 2.7   Line protection for a loop system. Equivalent sources on both ends of the line contribute current for a fault on the line. (a) System diagram. (b) Voltage and current phasors

protection, and operates when the fault is in the 'forward' direction – i.e. towards the zone of protection. These directional relays may use the phase angle between the fault current and some reference quantity (the corresponding voltage, for example) to determine the direction of the fault. Figure 2.7(b) shows the voltage behind the fault and the fault current. Assuming that the fault circuit consisting mostly of transmission lines is almost purely inductive the fault current lags the voltage by almost 90°. If the angle between the reference phasor (voltage) and the fault current is θ, then the relay operating principle could be described by

$$-\pi \leq \theta \leq 0, \text{operate}$$

$$0 \leq \theta \leq \pi, \text{block} \tag{2.2}$$

It should be noted that, although the region of operation and blocking takes up the entire plane ($0 \leq \theta \leq \pi$) in the above definition, the actual angles for a realistic fault will be around $-\pi/2$ for forward faults and around $\pi/2$ for reverse faults. Thus the operating principle could be made more selective by defining zones around $\pm\pi/2$. The voltage used as a reference must of course be the voltage that is driving the fault current. Thus for a phase a to ground fault, the a phase voltage and current must be used in this comparison. For a phase b-c fault, phase b current and voltage between phases b and c must be used. Occasionally, a current may be used as a reference phasor. Consider the transformer connected at the bus where the line under consideration originates (see Figure 2.8). For a ground fault within the zone of protection (e.g. at F_1), the transformer neutral current and the fault current will be in phase with each other. On the other hand, for a fault such as at F_2 the faulted phase current flowing through the relay under consideration will reverse, while the transformer neutral current will maintain its direction. Thus the transformer neutral current provides an effective reference for directional relays.

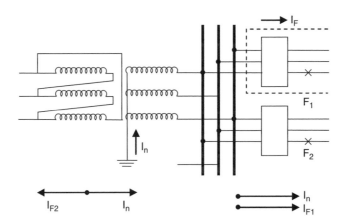

Figure 2.8   Use of transformer neutral current for polarizing a directional relay

The quantity providing the reference is often called a 'polarizing' quantity of the directional relay. The advantage of using current for polarizing is that the relay may be applicable where only CTs are available. The reader should beware that in case of an auto-transformer, the neutral current is not always dependable as a polarizing quantity. Instead, the current in a delta winding (if it is available) is far more desirable, although even this may be subject to reversals under certain conditions.

Generally, a protection system designed for phase faults is distinct from protection for ground faults. This is so because (a) ground fault currents are dependent upon system grounding, and (b) ground faults produce zero-sequence currents whereas there is very little zero-sequence current during normal operation. Thus the pickup settings of the ground fault relays can be made more sensitive than those of phase fault relays. It should be clear that the three phase-to-phase fault relays responding to the so-called delta currents, viz. $(I_a - I_b)$, $(I_b - I_c)$, $(I_c - I_a)$, are needed to protect against all phase-to-phase faults; while a separate relay responding to the zero sequence current. $(I_a + I_b + I_c)/3$ is provided for ground fault protection. Appropriate polarizing sources must also be provided for each relay if a directional overcurrent function is needed. Often one of the phase-to-phase relays may be omitted, as the remaining two phase-to-phase relays and the zero-sequence current relay provide adequate protection against all faults.

2.3.3 Distance relays

As mentioned in Section 2.3, the pickup value of an overcurrent relay must be set between the maximum load current and the minimum fault current experienced by the relay. In high voltage and extra-high voltage networks, these parameters are often not well defined, nor are they separated sufficiently from each other to allow a safe selection for a pickup setting. For such cases, the distance relay furnishes excellent protection under all circumstances. Consider the transmission line shown in Figure 2.9(a). Let there be a fault at a fractional distance k from the relay location. If there is a phase-to-phase fault between phases x and y such that x ≠ y; and x,y = a,b,c, then it can be shown that[6,7]

$$\frac{E_x - E_y}{I_x - I_y} = kZ_1 \tag{2.3}$$

where Z_1 is the positive sequence impedance of the entire line. Similarly, for a phase to ground fault on phase x

$$\frac{E_x}{I_x + mI_0} = kZ_1 \tag{2.4}$$

m being equal to $(Z_0 - Z_1)/Z_1$, and Z_0 is the zero sequence impedance of the line.[8] The ratios of appropriate voltages and currents represent the fraction of line

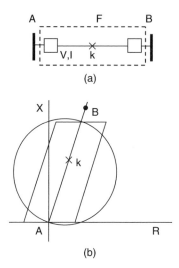

Figure 2.9 Distance relaying of high voltage transmission lines. (a) Line with a fault in the zone of protection. (b) Distance relay characteristic

impedance (positive sequence) at which a fault occurs. The computed ratio can be compared with the total positive sequence impedance of the zone being protected, and if smaller, trip output is produced. It must be noted that the ratio of the two phasors – numerator and denominator in Equations (2.3) and (2.4) being complex numbers is a complex number. Consequently, the comparison is made in the complex impedance plane as shown in Figure 2.9(b). For faults on the transmission line, the ratio is a complex number lying on A-B. However, allowing for various inaccuracies within the transducers and in the relay, as well as due to fault arc resistance, it becomes necessary to define the fault region in the complex plane as having a substantial area surrounding the line AB. A rectangle, circle, or a segment of a circle are all acceptable shapes, and define the zone of protection in the R-X plane. The circular shape originated with electromechanical relays and has been carried over in many solid-state relays (and in some computer relays) as well. The circular zone shown in Figure 2.9(b) belongs to the class of relays known as offset impedance relays – the center of the circle being offset from the origin. When the circle passes through the origin as illustrated, the shape is known as a 'mho' relay. Knowing the inaccuracies and fault resistance that must be allowed for, a more accurate zone shape can be defined so as to occupy a minimum area of the complex R-X plane. The rectangle (or more generally a quadrilateral) enclosing the line AB is a more appropriate shape for distance relaying, and most computer relays provide such a shape.

As in the case of overcurrent relays, complete protection with distance relays involves three phase distance relays (using delta voltages and currents) and three ground distance relays to protect against all possible faults. Since overcurrent relays can be used for ground fault protection effectively (there being no significant zero-sequence component in load currents), often three phase distance relays and one ground overcurrent relay provide adequate protection.

Yet another application related subject is the manner in which distance relays are put to use. The performance of a distance relay near its zone boundaries is not very predictable because of various types of errors mentioned earlier. Consequently, it becomes necessary to use multiple zones of protection to cover the entire line dependably and securely. Consider the protection of line AB in Figure 2.10(a). The zone of protection is that shown by a dotted line. However, to be sure of covering it in the presence of input errors, two zones (Zone 1 and Zone 2) are used. Zone 1 relay operates instantaneously (no intentional delay – i.e. in about one to two cycles) while a fault in Zone 2 causes the relay to operate with an added delay (generally of the order of 20 to 30 cycles). In this fashion, the entire line is protected even where the zone boundary is not very precisely determined. The Zone 2 operating delay is to permit other relays such as those belonging to lines BC and BD to operate for faults within their respective first zones – such as F_2 or F_3 – which may lie in Zone 2 of the relay protecting line AB. Remembering that a similar protection system exists at the B terminal looking towards A, it is clear that such a line protection scheme would provide high speed protection from both

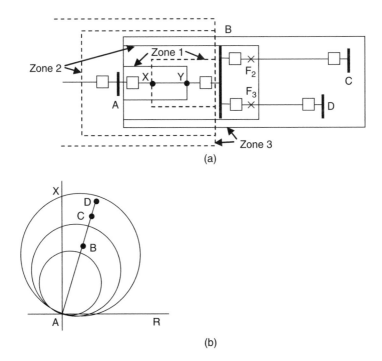

(a)

(b)

Figure 2.10   Step distance relaying of a transmission line. (a) Transmission lines protected by distance relays. (b) Three zones of protection

ends against faults in the middle portion of the line (in region XY); while faults on the line but outside the XY region are cleared instantaneously by the near relay and with a Zone 2 time delay by the distant relay. In addition to these two zones, often a third zone (with an additional time delay of the order of one second) is provided at each end in order to provide remote backup for the protection of the neighboring circuits. Zone 3 overlaps the longest line connected to the same bus as the line being protected. The three protection zones are shown in Figure 2.10(b). It should be noted that often, because of the system load, it is not possible to obtain a secure Zone 3 setting on high voltage or extra-high voltage networks.

2.3.4 Phasor diagrams and R-X diagrams[5]

The relay characteristics of Figures 2.9 and 2.10 are in the R-X plane. These diagrams are generally known as R-X diagrams, and are very useful in determining response of distance relays for different types of system conditions including faults, load changes, and power swings. It is well to understand how distance relay responses in the R-X plane can be derived from the phasor diagrams representing the current and voltage phasors representing the inputs to a distance relay.

Assume that inputs to the distance relay are voltage and current having phasor representations E and I. The key concept is that of 'apparent impedance' seen by the relay. Apparent impedance is defined as the ratio of the voltage phasor to the current phasor:

$$Z_{app} \equiv R + jX = E/I \tag{2.5}$$

Consider the case of the current phasor being $(1 + j0)$. In that case, the voltage phasor becomes equal to the apparent impedance. This concept can be formalized by a process of rotating the entire E-I phasor diagram until the current phasor becomes aligned with the real axis, and then changing the length of the phasors until the current phasor falls becomes 1.0 in length. The process is illustrated in Figure 2.11(a). The complex plane is now labeled R-X plane, and the apparent impedance seen by a relay supplied with phasor E and I is as shown.

Figure 2.11(b) illustrates the effect of variations in current magnitude and phase angle on the R-X diagram. For lagging currents, which correspond to active and reactive power flowing from the bus into the line (with current and voltage phasor reference directions as shown) the apparent impedance falls in the first quadrant. The right half of the R-X plane corresponds to real power flowing into the line, while the left half corresponds to real power flowing into the bus. Similarly the upper half of the R-X plane corresponds to reactive power flowing into the line, while the lower half corresponds to reactive power flowing into the bus. Increasing current magnitude or decreasing voltage magnitude brings the apparent impedance closer to the origin of the R-X diagram. This observation is at the heart of loadability limits of distance relays. As the apparent impedance approaches the origin of the

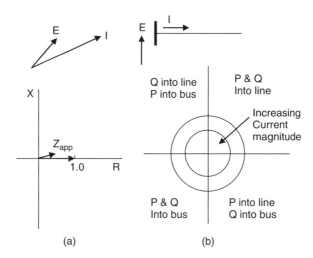

Figure 2.11 Apparent impedance seen by a relay energized by E and I phasors. (a) Conversion of phasor diagram to R-X diagram. (b) Apparent impedance behavior for different power system conditions

R-X diagram, it may eventually enter a characteristic of a relay located at the line terminal in question. Excursions by apparent impedance into the relay characteristics during load changes or power swings have caused unnecessary trips of relays, often contributing to cascading failures of the power systems.

In considering adaptive relaying applications in Chapter 10 we will make use of these concepts in order to understand the response of distance relays to various system conditions.

2.3.5 Pilot relaying

Pilot relaying is used to protect transmission lines when it is desired that an entire line (not just the region XY in Figure 2.10) be provided with high speed protection. This is particularly desirable in an integrated network, where with only the near breaker opening at high speed in zone 1 time, the delayed tripping of the remote breaker would be intolerable to the system. In a sense, the power system is so tightly knit that no fault is a 'distant' fault for which slow clearing can be accepted. Two types of pilot relaying systems are in general use: the directional comparison system and the phase comparison system. The actual implementation of each of these systems could lead to additional sub-classifications: for example, permissive or non-permissive, over or under- reaching.[2] The details of these schemes are not too important for the present discussion – although undoubtedly they must be taken into account when actual relay implementation is considered. The directional comparison scheme calls for communicating from one end of a line to the other whether a fault is in the direction of the zone of protection or in the opposite direction. As far as the relay itself is concerned, this direction determination is accomplished through a

directional distance calculation. Thus from the point of view of relaying algorithm
design, no new concept is introduced.

A phase comparison system is closely related to the differential protection principle. The relative phase angles of currents at the terminals of a line are compared
to determine if the algebraic sum of all currents entering a transmission line can
be non-zero. A true differential relay would require a check of the magnitude of
the algebraic sum as well. However, the communication channel requirements can
be kept at a modest level if only the phase information is exchanged. A segregated
phase comparison scheme applies the phase comparison criterion to each of the
three phase currents, whereas a combined phase comparison system uses a single
AC quantity derived from all three phase currents for the sake of such a comparison.
Clearly in the latter case, the communication channel requirements are substantially
reduced. A phase comparison relaying system is particularly suitable in case of
transmission lines with series capacitor compensation. Also, since the phase comparison is made for currents alone, no voltage input is needed for such systems.
On the other hand, directional comparison systems do require voltage inputs since
an impedance calculation is involved. It should be noted that, if the communication system should fail, the phase comparison scheme becomes totally inoperative,
while the directional comparison system can be designed to provide some distance
relaying functions as a second (additional) protective function.

2.4   Transformer, reactor and generator protection

2.4.1   Transformer protection

Small transformers are usually protected by fuses or overcurrent relays. Larger
transformers (2.5 MVA or greater in capacity) are usually protected by percentage current differential relays. Consider a two-winding single-phase transformer
illustrated in Figure 2.12. When the transformer is without a fault within the zone
defined by the two CTs

$$I_1 N_1 = -I_2 N_2 T \tag{2.6}$$

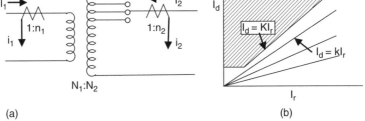

(a)   (b)

Figure 2.12   Percentage differential protection of a transformer. (a) The single-phase transformer. (b) Slope of the percentage differential characteristic

Equation (2.6) is an approximation, because it does not take into account the magnetizing current. N_1 and N_2 are the nominal turns of the two windings, and T is the ratio of the tap changer. If the two current transformers have turns ratios of 1:n_1 and 1:n_2, respectively, then

$$I_1 = n_1 i_1$$

$$I_2 = n_2 i_2 \tag{2.7}$$

When the tap changer is at the neutral tap setting (i.e. when $T = 1$), the CT secondary currents i_1 and i_2 may be made equal in magnitude by choosing n_1 and n_2 such that

$$N_1 n_1 = N_2 n_2 \tag{2.8}$$

Since the current transformers are selected from available standard ratio CTs, in general $N_1 n_1 \neq N_2 n_2$, and $i_1 - i_2 \neq 0$ for a transformer without a fault. The tap changer creates an additional disparity between i_1 and i_2 when it deviates from its nominal value. And finally, the CT errors also make a contribution to the algebraic sum of i_1 and i_2. In general then,

$$i_1 + i_2 = k \left[\frac{i_1 - i_2}{2} \right] \tag{2.9}$$

or

$$I_d = k I_r$$

The algebraic sum $(i_1 + i_2)$ is the differential current I_d, and $(|i_1| + |i_2|)/2$ is the average value of the two winding currents referred to the CT secondaries. This is known as the restraining current I_r, and Equation (2.9) indicates that, when the transformer is without an internal fault, a differential current equal to k times the restraining current may be developed. For the differential relay to refrain from tripping, it becomes necessary to shape the relay characteristic as shown in Figure 2.12(b). The percentage slope K in Figure 2.12(b) is made greater than the k of Equation (2.9), in order to allow for some safety margin. The constant k has three contributing factors as shown in the Figure. The smaller the setting K of the relay, the more sensitive is the relay in detecting small fault currents. Typical settings available for percentage differential relays are 10, 20 or 40%.

During energization of a transformer, abnormal currents may flow in the winding that is being energized. These are known as the magnetizing inrush currents, caused by the saturation of the transformer core for portions of a cycle. Typical inrush currents are shown in Figure 2.13. The inrush may be quite severe if there is remanence in the core and it is of a polarity which takes the core further into saturation. As the relationship between the remanence and the flux build up caused by energization is random, actual inrush obtained during an energization depends upon chance. However, the fact remains that, during energization, high current

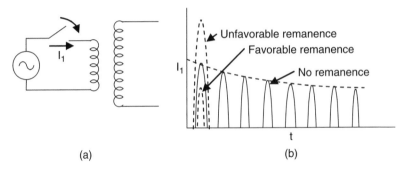

Figure 2.13 Inrush current during energization of a transformer. The amount of inrush depends upon the instant of switching and upon the remanence in the core

may flow in the primary of a transformer. This is exactly the condition obtained when there is an internal fault in the transformer. It is thus necessary to distinguish between a fault and an inrush transient.

The accepted technique for blocking the trip action of a percentage differential relay under conditions of inrush is to make use of the high second harmonic current in the inrush, whereas a fault current is almost purely of fundamental frequency. The second harmonic current is used as another restraining signal (in addition to the fundamental frequency restraining current I_r of Equation (2.9)). In designing a harmonic restraint percentage differential relay, note must be taken of other phenomena which produce harmonics in the current waveforms.[9] For example, an overexcited transformer has significant fifth harmonic component in its magnetizing current, and hence it is desirable to create a composite restraint function with 2nd and 5th harmonics. On the other hand, if one of the CTs saturates during an internal fault, the resulting third harmonic current in the secondary winding of the saturated CT should not produce any restraint function.

Although current harmonics furnish a sound means of distinguishing faults from no-fault conditions, other schemes for achieving the same end result are possible. Thus a high voltage at the transformer terminal may also be used to indicate that any differential current present must be due to magnetizing inrush. This, and some other ideas, will be discussed more fully in Chapter 5.

It should also be noted that some of the most sensitive relays for transformer protection are not electrical in nature. These are the Sudden Pressure Relay (SPR) and the Buchholz relay.[1]

Three phase transformer protection generally follows the principles outlined above. In case of a wye-delta transformer, the line currents during normal operation (or, through currents flowing when there is an external fault) on the wye and the delta sides have a phase difference between them. This must be accounted for before a differential relay can be connected. This is usually accomplished by connecting the CTs in a reversed connection: in wye on the delta side of the main

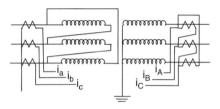

Figure 2.14 Wye-delta transformer protection by percentage differential relay. The CT connections compensate for the phase shift in the main transformer

transformer and in delta on the wye side. This is illustrated in Figure 2.14. The same effect may be achieved through computation in a computer relay.

2.4.2 Reactor protection

The main protection for a reactor is similar to generator differential protection, discussed next. In addition, a turn-to-turn fault in a reactor can be detected by a distance relay connected to look into the reactor.

2.4.3 Generator protection

The current transformers used at the two ends (high side and neutral) of a generator winding are specially matched so as to reduce the disparity in their performance. This is possible in the case of generators (and reactors) because the primary currents of the two CTs are identical (as was not the case for a transformer differential relay). Also no allowance need be made for errors caused by tap changers. Thus, a generator differential protection relay can be made extremely sensitive. Of course, there is no need to worry about the magnetizing inrush current.

Another concern in case of generator protection is the rotor heating caused by unbalanced stator currents. This is usually related to the amount of heating of the rotor caused by negative sequence current in the stator windings. The negative sequence current relay tests for the quantity

$$\int i_2^2 \, dt \geq K, \text{ trip or alarm,}$$

$$\leq K, \text{ no operation.} \tag{2.10}$$

The constant K is usually furnished by the generator manufacturer. Several other generator relaying tasks are closer to being control tasks in that operator intervention or automatic control actions (slower than relay responses) are called for. Examples of these functions are reverse power relay, field ground relay, generator capability (overload) relay, and out-of-step relays. In Chapter 5 we will consider implementation of some of these functions with the help of conventional relays described in Section 2.2.

2.5   Bus protection

A bus, being a power system element that does not extend over long distances (as transmission lines do), is ideally suited for protection by a differential relay. Consider a bus and its associated circuits consisting of lines or transformers as shown in Figure 2.15. The algebraic sum of all the circuit currents must be zero when there is no bus fault. With all circuit CT ratios being equal, the secondary currents also add to zero when there is no bus fault. The various CT inaccuracies require that a percentage differential relay be used, but in this case the percentage slope can be quite small, as there are no mismatched ratios or tap changers to be concerned about. And of course, there is no magnetizing inrush phenomenon to consider.

One area of concern is the saturation of a CT during an external fault. Consider the fault at F in Figure 2.15. The current in the CT on this feeder is the sum of all feeder currents, and consequently this CT is in danger of becoming saturated. A saturated CT produces no secondary current while the CT core is in saturation.[10] Typical waveforms of primary current, secondary current and the flux in the CT core are given in Figure 2.16. Whenever the flux density crosses the saturation level,

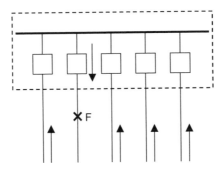

Figure 2.15   Bus protection with a differential relay. When there is no fault, the algebraic sum of circuit currents is zero

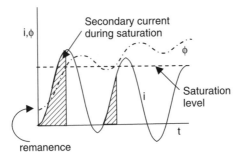

Figure 2.16   Primary current, secondary current, and the flux in the core of a CT in saturation. Remanence may cause saturation to set in earlier and last longer

the secondary current becomes negligible. Under these conditions, the secondary winding is no longer strongly coupled with the primary winding – the transformer essentially acts like an air-core device.

A lack of strong coupling implies that the secondary winding presents a very low impedance to any external circuit connected at its terminals, instead of acting like a current source of high equivalent impedance. It should be clear that, if the secondary current in one CT becomes zero for any period during an external fault, the differential current will be equal to the missing current causing the relay to trip. In general the core of a properly chosen CT should not saturate within 1/2 to 1 cycle of fault inception. However, often the requirement placed on bus differential relays is that they should restrain from operating for external faults even if a CT should saturate in 1/4 cycle or less after the occurrence of a fault. This requirement places a very confining restriction on a computer based bus differential relay. This subject will be dealt with in Chapter 6.

However, analog relays have a very ingenious solution to the problem posed by a saturating CT. Since the saturated CT secondary appears as a low impedance path in the differential circuit, it is sufficient to make the relay a high-impedance device. The spurious differential current produced by the saturated CT then flows through its own secondary winding and by passes the relay having a much higher impedance. The condition is illustrated in Figure 2.17. The saturation of the CT is itself responsible for saving a false operation which would have resulted from the saturation.

2.6 Performance of current and voltage transformers

2.6.1 Current transformers

We have already discussed in Section 2.5 the CT secondary current error caused by saturation of the CT cores. These are quite substantial – when they occur the secondary current disappears for a portion of the waveform. However, even when the CT core is not saturated to such an extent, the secondary current is always in error due to the small but non-zero magnetizing current required to set up the

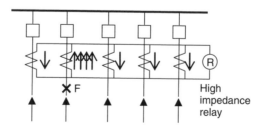

Figure 2.17 Effect of saturating CT on a high impedance differential relay. The lower impedance of the saturated CT secondary bypasses the differential current from the relay

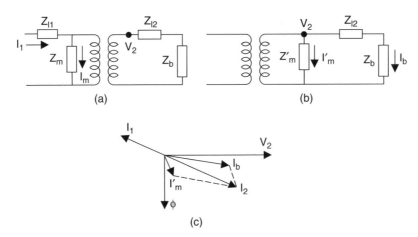

Figure 2.18 (a) Complete equivalent circuit of a current transformer. (b) A useful approximation for practical CTs. (c) Phasor diagram

flux in the core. Consider the equivalent circuit of a current transformer, shown in Figure 2.18(a). The leakage reactance of primary and secondary windings is generally negligible in most practical current transformers; hence the equivalent circuit of Figure 2.18(b) is more appropriate for its analysis. The load (burden) impedance Z_b includes resistance of all the secondary wiring as well as the impedance of all instruments and relays connected to the CT. The primary current I_1, the secondary current I_b, the magnetizing current I'_m, the core flux ϕ, and the voltage across the secondary winding V_2 are shown in the phasor diagram of Figure 2.18(c). Note that, although the phasor diagram is generally used in this analysis, there may be some harmonics present in the magnetizing current due to the nonlinearity of the core B-H characteristic. In any case, I_2 differs from I_b by I'_m, which is therefore the CT error for the primary current and burden impedance shown. It should be clear that a smaller burden impedance Z_b causes smaller V_2 and I'_m; thereby reducing the CT error. If the magnetizing characteristic of the core is available, the CT error can be evaluated for any operating conditions of the CT. The CT classifications (such as 10C400 or 10T400) indicate the upper limit of the CT error as being 10% when the primary current is 20 times normal and the secondary voltage is less than or equal to 400 volts. The letters C or T in the class designation indicate the type of CT construction – C type being a better CT than the T type. The CT error is generally kept low by a proper choice of CT and its connected burden. The only point of departure for computer relaying is that conceivably the CT error could be computed and corrected inside the computer relay if the CT characteristic and burden impedance were given as inputs to the computer relay. This clearly cannot be done in conventional relays. This idea has not been used in computer relaying applications developed so far.

2.6.2 Voltage transformers

Some voltage transformers, especially those for lower voltage transmission, are magnetically coupled potential transformers with a primary winding and a secondary winding. Such transformers are very accurate, and in general their transformation errors can be neglected.

A fairly common voltage transformer uses a capacitor voltage divider network, as shown in Figure 2.19(a). The voltage divider reduces the line potential to a few kV, and is further reduced to the standard relaying voltage of 67 volts line-to-neutral by a magnetic core transformer. The capacitive voltage divider presents a capacitive Thévenin impedance as shown in Figure 2.19(b), and in order to eliminate any phase angle errors due to the load current flowing through the capacitive Thévenin impedance, a tuning inductance L is connected in series with the primary winding. By making $1/2\pi f(c_1 + c_2) = 2\pi fL$, the phase shift across $(c_1 + c_2)$ is exactly canceled by the phase shift across L at all load currents, and once again the secondary voltage is in phase with the primary voltage. In general, the steady state error of the Capacitive Voltage Transformer (CVT) is negligible.

However the transient response of a CVT is of some concern in relay design. The equivalent circuit of Figure 2.19(b) can be used to determine the output waveform across the burden when a fault occurs on the primary circuit. As the primary voltage changes suddenly from its pre-fault value to its (smaller) post-fault value, the output voltage undergoes a subsidence transient before settling to its final steady state value. The subsidence transient magnitude depends upon CVT parameters, burden impedance and power factor, and upon the angle of incidence of the primary fault.

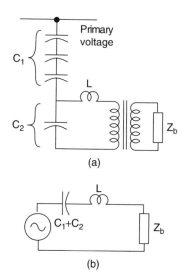

Figure 2.19   (a) Capacitive voltage transformer. (b) Equivalent circuit

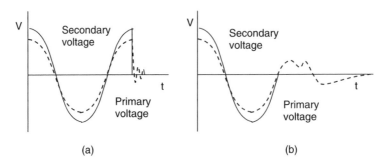

Figure 2.20 CVT transient response. (a) Complete primary voltage collapse at voltage maximum. (b) Voltage collapse at zero voltage. Note that the secondary transient is much more pronounced in case (b)

In general, the faults occurring at or near zero voltage produce the worst subsidence transient. Figures 2.20(a) and (b) show samples of CVT response for representative installations.[11] Although omitted from most simulation studies, usually a ferroresonance suppression circuit is also present on the CVT secondary side, and does affect the transient response.

The CVT transient response causes difficulties in those relaying tasks which require voltage inputs. In particular, faults causing near-complete voltage collapse create false voltage pictures at the relay input terminals. Special attention should be given to relay algorithm design in such cases, particularly if severe voltage collapse may be caused by a fault near a zone boundary. Short transmission lines fed from weak systems usually constitute difficult relay design problems due to transient CVT errors.

2.6.3 Electronic current and voltage transformers

Electronic current transformers were discussed in technical literature in the early 1960s, but their practical realization useable in modern power systems became possible in the 1980s. In recent years there have been developments of electronic current and voltage transformers which are promising source of input signals for computer based relays. We will present a brief account of their principles of operation here and refer the reader to the references cited above for additional information.

2.6.3.1 Electronic current transformers[12]

The operating principle of these devices (also known as Magneto-Optic Current Transformers) is based on Faraday Effect, by which the plane of polarization of a polarized light beam is deflected by an angle which depends upon the magnetic field to which the beam is subjected.

The principle of operation of electronic current transformers is shown in Figure 2.21. A light beam is produced by a LED, and collated by a lens before

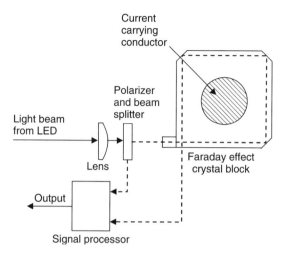

Figure 2.21   Schematic of the principle of operation of an electronic current transformer

it passes through a polarizer, where the beam is split into two parts. One part of the beam goes directly to a detector in the signal processing circuit, while the second part of the linearly polarized light beam is passed through a special crystal block with total reflections taking place at its three corners so that the light beam makes a complete circuit around the conductor through which the current to be measured is flowing. By Ampere's law, a complete circuit through the magnetic field produced by the current leads to a closed path integral of the magnetic field, which is proportional to the current. The deflection of the plane of polarization of the light is thus proportional to the instantaneous value of the current. The beam with deflected plane of polarization is also sent to the signal processor, where the angle of rotation of the beam as it went through the magnetic field is measured by comparison with the first half of the beam, and an output voltage signal proportional to the instantaneous current value is produced. This signal can then be routed to appropriate application. It should be clear that the voltage signal is easily handled by computer relays, while traditional relays requiring heavy current inputs must be supplied through a current amplifier.

2.6.3.2   Electronic voltage transformers[13]

The electronic voltage transformer is based upon the electro-optic device known as Pockels cell, and is illustrated in Figure 2.22. A light beam produced by a LED and collated by a lens, as in the case of the electronic current transformer is passed through a polarizer and a quarter-wave plate, producing a circularly polarized light beam. This beam is passed through a Pockels cell which is subjected to an electric potential field produced by applying a voltage in a direction perpendicular to the direction of the light beam. The effect of the passage through the electric field on

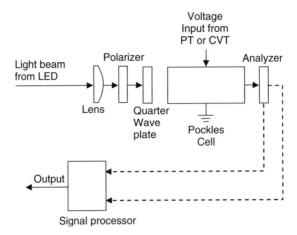

Figure 2.22   Schematic of the principle of operation of an electronic voltage transformer

the light beam is to convert the circularly polarized light beam to an elliptically polarized light beam, with the degree of ellipticity being proportional to the strength of the electric field. An analyzer splits the elliptically polarized beam in two linearly polarized beams, with their planes of polarization perpendicular to each other. The relative intensity of each of the beams is compared in the signal processor, which measures the degree of ellipticity which is proportional to the instantaneous electric field (and by inference the instantaneous value of the applied voltage). As before, the measurement is converted to a voltage which reproduces the voltage applied to the Pockels cell.

2.6.3.3   Rogowski coils

Rogowski coil is a helical coil placed in a loop around the current carrying conductor with a return conductor placed in the center of the helical coil as illustrated in Figure 2.23. The coil has a non-magnetic core (for example air), and thus has no saturation effects. The coil has a voltage induced which is proportional to the rate

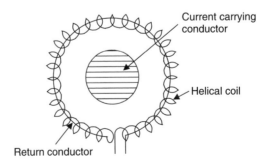

Figure 2.23   Schematic of the principle of operation of a Rogowski coil

of change of its flux linkage. Since the flux is proportional to the current in the conductor in the center, the voltage in the Rogowski coil is proportional to the rate of change of current. This signal can be processed through an integrator (or first sampled and then integrated digitally) to provide a signal proportional to the current. Clearly the output of such a system is not suitable for conventional relays which require large current signals to operate, but is convenient for computer relays.

The return conductor passing through the center of the coil cancels any induced voltage due to any other interfering magnetic field which may be present. Although this is not a main-stream current measuring circuit, versions of this design have been used in integrated sensing and protection applications.[14]

2.7 Summary

In this chapter, we have presented a classification of relay types, and have described their behavior with the help of an input-output relationship. We have pointed out the importance of the concept of a phasor during the occurrence of transient phenomena. We have given a short account of some of the relay application considerations; however, we have accepted that a thorough discussion of relay applications is beyond the scope of this book. Finally, we have pointed out the causes of errors created by current and voltage transformers. These errors ultimately affect the accuracy with which a relay can discriminate between faults internal to its zone of protection, and all other system transients. We have also described electronic current and voltage transformer principles, which eliminate some of these sources of errors. This very promising technology is still not in wide use in power systems. In the following chapter we will begin a discussion of the mathematical basis for relaying algorithms, and then take up an account of relaying algorithms which are in use at the present time.

Problems

2.1 Fault calculations are usually performed with symmetrical components. Review the procedure for calculating faults on a three-phase network with the help of the sequence diagram. Verify that a three-phase fault involves only the positive sequence network, the phase-to-phase fault involves the positive and the negative sequence networks, and that all ground faults require the zero sequence network, as well. Show that the phase-to-ground fault on phase a is represented by a series connection of the sequence networks at the point of fault. What would be the corresponding connection for a phase b to ground fault?

2.2 A radial feeder is connected to a source having an impedance of 0.5 per unit. The feeder length is 10 miles with a positive sequence reactance of 0.7 ohm per mile, and a zero sequence impedance of three times that value. Calculate the three-phase fault current, phase-to-phase fault current, and the phase-to-ground fault current as a function of the fault location along the feeder. Plot the three fault current magnitudes as a function of the distance from the sending end. Assume that the zero sequence impedance of the source is equal to its positive sequence impedance.

2.3 The overcurrent relay at bus A in Figure 2.4 is to be coordinated with the relay to the right of bus B. The time dial setting of the relay at B is 1/2. It picks up reliably for a minimum fault current of 650 amperes. Assuming a CT ratio for the relay at bus B of 200:5, what should be the relay tap setting of this relay if any desirable tap setting could be available? The maximum fault current for a fault at bus B is 1000 amperes. Determine the time dial setting, CT ratio and the relay tap setting for the relay at A. Assume the relay characteristics as shown in Figure 2.6.

2.4 In problem 2.3, what is the safe load current (loadability) that can be carried by the two relays?

2.5 Consider a three-phase autotransformer. Assume the primary voltage to be 69 kV, while the secondary voltage is 345 kV. The transformer leakage reactance is 0.1 per unit on its own base. Assume a phase-to-ground fault at varying distance from the terminals of the transformer (i.e. assume a variable impedance between the transformer terminals and the fault point). Show that the current in the neutral of the autotransformer is not always flowing up the neutral, indicating that this current is an unsuitable polarizing current for directional ground relaying.

2.6 Using the symmetrical component representation for phase-to-phase and phase-to-ground faults developed in problem 2.1, show that Equations (2.3) and (2.4) are valid.

2.7 A 'mho' characteristic is a circle passing through the origin in the complex R-X plane. Assuming the diameter of the circle to be $(r + jx)$, derive an equation for the voltages and currents seen by a relay such that the voltage-current pairs correspond to points on the relay characteristic.

2.8 A single-phase transformer is rated at 34.5 kV primary and 345 kV secondary. The transformer is rated at 50 MVA. Determine the CT ratios to be used for differential relaying of this transformer. Using standard types of CTs, what should be the slope of the percentage differential relay if the transformer has a tap changer with a range of ±6%? Allow for CT errors as well as a safety margin for security.

2.9 Assume a normal load current flow in a three-phase wye-delta transformer. Show the connections of a differential relay and the CT secondary windings so that the differential relay is in balance (i.e. no differential current is produced for this condition). Assume suitable transformation ratios for the main transformer and the current transformers.

2.10 If the B-H curve of the core steel of a transformer consists of a vertical line when the core is unsaturated, and a line of some finite slope when the core is saturated, the inrush current is made up of portions of a sinusoid that are symmetric around its peak, and have a span of angle θ where $0 < \theta < 2\pi$. Determine the harmonic content of such a waveform for a general θ. Confirm that the second harmonic predominates, and that the harmonic magnitudes go to zero as θ goes to 0 or 2π.

2.11 For a B-H curve for the core steel as described in Problem 2.10, determine the shape of magnetizing current as the transformer is overexcited. Assume that over-excitation produces a symmetric flux excursion beyond the saturation level by a factor k, where k is greater than 1. Determine the harmonic content of the resulting magnetizing current, and show that fifth is the predominant harmonic in this case.

2.12 Assume the primary voltage of the CVT shown in Figure 2.18 to be 400 kV, and the voltage produced by the capacitive potential divider to be 4 kV. Assume the capacitor C_2 to be 0.1 μF. Determine the value of the tuning inductance. If the burden is 200 ohms resistive, determine the subsidence transient at the burden when the primary voltage collapses to zero at (a) peak of the voltage wave, and (b) at the zero-crossing of the voltage wave.

References

[1] Mason, C. R. (1956) *The Art and Science of Protective Relaying*, John Wiley & Sons, Ltd.

[2] *Applied Protective Relaying*, Westinghouse Electric Corporation, 1982.

[3] Van C. Warrington, A. R. (1962, 1969) Protective Relays, Their Theory and Practice, vol. I, Chapman and Hall, London and John Wiley & Sons Inc. New York, 1962; vol. II, Chapman and Hall, London and John Wiley & Sons Inc. New York, 1969.

[4] *Power System Protection*, vols. I, II, III. Edited by Electricity Council. The Institution of Electrical Engineers.

[5] Horowitz, S. H. and Phadke (2008) A.G. *Power System Relaying*, Third edition, John Wiley & Sons, Ltd.

[6] Horowitz, S. H. (ed.) (1980) *Protective Relaying for Power Systems*, IEEE Press, New York.

[7] Stevenson, Jr., W. D. (1980) *Elements of Power System Analysis*, Chapter 13, McGraw-Hill, New York.

[8] Lewis W. A. and Tippett, L. S. (1947) Fundamental basis for distance relaying on 3-phase systems, AIEE Transactions, vol. 66, pp. 694–708.

[9] Einvall, C. H. and Linders, J. R. (1975) A three-phase differential relay for transformer protection, IEEE Trans. on PAS, vol. PAS-94, no. 6.

[10] IEEE Publication no. 6 CH1130-PWR, *Transient Response of Current Transformers*, 1976.

[11] Sweetana, A. (1970) Transient response characteristics of capacitance potential devices, *IEEE Trans. on PAS*, Vol. PAS-89, pp. 1989–1997.

[12] Emerging Technologies Working Group and Fiber Optic Sensors Working Group, Optical current transducers for power systems: a review, IEEE Transactions on Power Delivery, vol. 9, no. 4, October 1994, pp. 1778–1788.

[13] Kanoi M., Takahashi, G., Sato, T. *et al*. (1986) Optical voltage and current measuring system for electric power systems, IEEE Transactions on Power Delivery, vol. PWRD-1, no. 1, pp. 91–97.

[14] Ljubomir K. (2002) PCB Rogowski coils benefit relay protection, IEEE Computer Applications in Power, July 2002.

3

Mathematical basis for protective relaying algorithms

3.1 Introduction

One of the attractions of the computer relaying area is the rich combination of academic disciplines upon which it is based. In addition to protective relaying itself, computer relaying depends on computer engineering for an understanding of hardware selection and compromises, on communication systems for an understanding of the links between devices, and on elements of digital signal processing and estimation theory in understanding the algorithms. To understand a number of algorithms it is necessary to understand how the Fourier series and Fourier transform are used to extract the fundamental frequency component of voltage and current waveforms. The first few sections of this chapter are devoted to a brief presentation of Fourier series and transform ideas along with a related expansion of a signal in terms of Walsh functions. Another important concept in evaluating relay algorithms is that of estimation. Since unwanted components are generally present in the signals that are sampled, it is necessary to estimate the parameters of interest (the fundamental frequency component, for example) by processing a number of samples. The remaining material in this chapter is concerned with ideas of probability, random processes, and estimation including the Kalman filter. Our treatment of all of these subjects is, of necessity, brief. The interested reader is referred to more complete books on these subjects given as references at the end of the chapter.

3.2 Fourier series

Many of the input signals such as phase voltages and currents encountered in power systems are essentially periodic. Ideally the voltages and currents present in the system in steady state are pure sinusoids at the power system frequency (50 or

Computer Relaying for Power Systems 2e   by A. G. Phadke and J. S. Thorp
© 2009 John Wiley & Sons, Ltd

60 Hz). Some devices (for example, power transformers, inverters, converters and loads) create harmonic distortion in the steady state signals. The signals seen by protective relays also fail to be pure sinusoids. The non-fundamental frequency content of the voltage and current seen by a relay are not truly periodic but change in time. The nature of these non-fundamental frequency signals has an important bearing on the performance of relaying algorithms. The Fourier series provides a technique for examining these signals and determining their harmonic content.

A signal r(t) is said to be periodic if there is a T such that

$$r(t) = r(t + T); \quad \text{for all } t \tag{3.1}$$

If r(t) is periodic and not a constant then let T_0 be the smallest positive value of T for which Equation (3.1) is satisfied. The period T_0 is called the fundamental period of r(t). The need for a concern with the smallest such T is made clear by considering a sinusoid. If $r(t) = \sin(\omega_0 t)$ then Equation (3.1) is satisfied for

$$T = \frac{2n\pi}{\omega_0}; \quad n = 1, 2, 3, \ldots \tag{3.2}$$

The smallest positive value is, of course, $T = 2\pi/\omega_0$. Associated with the fundamental period is a fundamental frequency defined by

$$\omega_0 = \frac{2\pi}{T_0} \tag{3.3}$$

Example 3.1

If
$$r(t) = e^{j\omega_0 t}$$

then Equation (3.1) requires
$$e^{j\omega_0(t+T)} = e^{j\omega_0 t}$$
$$e^{j\omega_0 t}\lfloor e^{j\omega_0 T} - 1 \rfloor = 0; \quad \text{for all } t$$

which implies that $\omega_0 T = 2n\pi$; n = 1, 2, 3, ... The conclusion is that the fundamental period is $T_0 = 2\pi/\omega_0$, and the fundamental frequency is ω_0.

Example 3.2

The real and imaginary parts of Example 3.1, viz. $\cos(\omega_0 t)$ and $\sin(\omega_0 t)$, have fundamental period $T_0 = 2\pi/\omega_0$ and fundamental frequency ω_0.

Example 3.3

The periodic square wave shown in Figure 3.1 (the dots indicate the signal is periodic) has fundamental period T_0 and fundamental frequency $\omega_0 = 2\pi/T_0$.

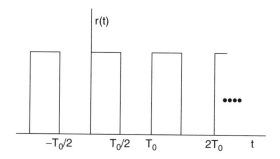

Figure 3.1   Periodic square wave

If $r_1(t)$ and $r_2(t)$ are both periodic signals with fundamental periods T_0 and T_1, respectively, then the sum, $r_1(t) + r_2(t)$, is not necessarily periodic. Consider

$$r(t) = \sin(t) + \sin(\pi t).$$

The fundamental period of $\sin(t)$ is 2π while the fundamental period of $\sin(\pi t)$ is 2. It is not possible to find integers n and m such that $2m = 2\pi n$ since π is irrational (not the ratio of integers) so that there is no T that satisfies Equation (3.1). On the other hand if

$$r(t) = e^{j\omega_0 t} + e^{2j\omega_0 t}$$

then the two fundamental periods differ only by a factor of two, i.e. $2\pi/\omega_0$ and π/ω_0. Thus the fundamental period of the sum is the larger, $2\pi/\omega_0$. In fact it is easy to see that any finite sum of the form

$$r(t) = c_0 + c_1 e^{j\omega_0 t} + c_2 e^{2j\omega_0 t} + c_3 e^{3j\omega_0 t} + \cdots + c_N e^{jN\omega_0 t}$$

is periodic with fundamental frequency ω_0. Including both positive and negative terms in the sum,

$$r(t) = \sum_{k=-N}^{k=N} c_k e^{jk\omega_0 t} \tag{3.4}$$

is also a periodic signal with fundamental frequency ω_0. The objective of Fourier analysis is to decompose an arbitrary periodic signal into components as in Equation (3.4).

3.2.1 Exponential fourier series

Given a periodic signal with fundamental frequency ω_0 the exponential Fourier series is written as[1]

$$r(t) = \sum_{k=-\infty}^{k=\infty} c_k e^{jk\omega_0 t} \qquad (3.5)$$

The task at hand is to determine the coefficients c_k. An important property of the exponentials makes the calculation a simple process. Note that

$$\int_0^{T_0} e^{jm\omega_0 t} \, dt = \begin{cases} T_0; & m = 0 \\ 0; & m \neq 0 \end{cases} \qquad (3.6)$$

The value of the integral is clear for $m = 0$, while for $m \neq 0$

$$\int_0^{T_0} e^{jm\omega_0 t} \, dt = \frac{1}{jm\omega_0} \left[e^{jm\omega_0 t} \right]_0^{T_0} = 0$$

since $\omega_0 T_0 = 2\pi$. Equation (3.6) is also true for integration over any period, i.e. the integrand could have been from τ to $\tau + T_0$. To compute the Fourier series coefficients it is only necessary to multiply Equation (3.5) by $e^{-jn\omega_0 t}$ and integrate over a period:

$$\int_0^{T_0} r(t) e^{-jn\omega_0 t} \, dt = \int_0^{T_0} \sum_{K=-\infty}^{K=\infty} c_k e^{j(k-n)\omega_0 t} dt \qquad (3.7)$$

From Equation (3.6) every term on the right hand side vanishes except the nth, yielding

$$\int_0^{T_0} r(t) e^{-jn\omega_0 t} \, dt = T_0 c_n \qquad (3.8)$$

or

$$c_n = \frac{1}{T_0} \int_0^{T_0} r(t) e^{jn\omega_0 t} \, dt \qquad (3.9)$$

Equation (3.9) can be evaluated over any period that is convenient as will be seen in example to follow.

Example 3.4

Consider the square wave of Figure 3.1. The nth coefficient is given by

$$c_n = \frac{1}{T_0} \int_0^{T_0/2} e^{-jn\omega_0 t} \, dt = \left[\frac{e^{-jn\omega_0 t}}{-jn\omega_0 T_0} \right]_0^{T_0/2}$$

$$= \left[\frac{e^{-jn\omega_0 T_0/2} - 1}{-jn\omega_0 T_0} \right]$$

$$= \left[\frac{e^{-jn\pi} - 1}{-jn2\pi} \right] = \frac{(-1)^n - 1}{-j2n\pi}$$

$$c_n = \frac{1}{jn\pi}; \, n \text{ odd}$$

$$c_n = 0; \quad n \text{ even}, n \neq 0$$

$$c_0 = 1/2$$

The coefficient c_0 (the average value of the signal) must frequently be evaluated separately. The approximations formed by the finite sum

$$r(t) = \sum_{k=-N}^{k=N} c_k e^{jk\omega_0 t}$$

for $N = 1, 3,$ and 5 are shown in Figure 3.2.

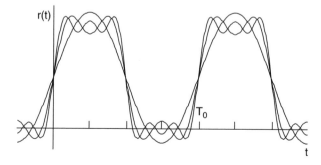

Figure 3.2 Finite Fourier series approximations to the square wave

3.2.2 Sine and cosine fourier series

Through the use of the Euler identity

$$e^{jk\omega_o t} = \cos(k\omega_o t) + j\sin(k\omega_o t) \tag{3.10}$$

it is possible to write the exponential series given in Equation (3.4) in terms of sines and cosines

$$\sum_{k=-\infty}^{\infty} c_k e^{jk\omega_o t} = a_o + \sum_{k=1}^{\infty} a_k \cos(k\omega_o t) + \sum_{k=1}^{\infty} b_k \sin(k\omega_o t) \tag{3.11}$$

where

$$a_o = c_o$$

$$a_k = c_k + c_{-k} \quad k \neq 0 \tag{3.12}$$

$$b_k = j(c_k - c_{-k}) \quad k \neq 0 \tag{3.13}$$

Or, using Equation (3.9),

$$a_k = \frac{2}{T_o} \int_0^{T_o} r(t) \cos(k\omega_o t)\, dt \tag{3.14}$$

$$b_k = \frac{2}{T_o} \int_0^{T_o} r(t) \sin(k\omega_o t)\, dt \tag{3.15}$$

From the results of Problem 3.2, it follows that real and even signals have expansions of the form of Equation (3.11) with only cosine terms, while real and odd signals have such expansions with only sine terms.

Equations (3.12) and (3.13) result from expanding the kth and $-k^{th}$ terms of the exponential series using the Euler expansion

$$c_{-k}e^{-jk\omega_o t} + \ldots\ldots + c_k e^{jk\omega_o t} = c_{-k}\cos(\omega_o t) + c_k \cos(\omega_o t)$$

$$-c_{-k}j\sin(\omega_o t) + c_k j\sin \omega_o t$$

If the Fourier series is written as

$$r(t) = c_o + (c_1 e^{j\omega_o t} + c_{-1}e^{-j\omega_o t}) + (c_2 e^{2j\omega_o t} + c_{-2}e^{2j\omega_o t}) \ldots$$

$$+ (c_k e^{jk\omega_o t} + c_{-k}e^{-jk\omega_o t}) + \ldots$$

the various terms can be recognized as

c_0 the dc component (average value)

$(c_1 e^{j\omega_0 t} + c_{-1} e^{-j\omega_0 t})$ the fundamental frequency component

$(c_k e^{jk\omega_0 t} + c_{-k} e^{-jk\omega_0 t})$ the kth harmonic

It is possible to think of the Fourier series expansion as the resolution of a periodic function into its frequency components. The frequencies present in a periodic function with fundamental frequency ω_0 are: 0(DC), $\pm\omega_0$ (the first harmonic), $2\omega_0$ (the second harmonic), $\pm 3\omega_0$ (the third harmonic), etc. The coefficient c_k then represents the amount of the signal at the frequency $k\omega_0$. The interpretation of the Fourier series coefficients as the frequency content of the signal can be summarized by plotting the magnitudes of the coefficients versus frequency.

Example 3.5

The half-wave rectified sinusoid shown in Figure 3.3 represents a first approximation to a current waveform encountered in power transformers under a condition of magnetizing inrush. The fundamental frequency is ω_0 and the Fourier series coefficients are given by

$$c_k = \frac{1}{T_0} \int_0^{T_0} r(t) e^{-jk\omega_0 t}\, dt$$

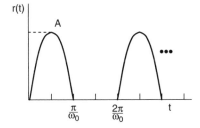

Figure 3.3   Half-wave rectified sine. An approximation to an inrush current

The coefficient c_1 is found to be $A/4j$ while all other odd c_k are zero and

$$c_k = \frac{A}{\pi(1 - k^2)} : k \text{ even}$$

Hence the first few terms of the Fourier expansion are

$$r(t) = \frac{A}{\pi} + \frac{A}{2} \sin(\omega_0 t) - \frac{2A}{3\pi} \cos(2\omega_0 t) - \frac{2A}{15\pi} \cos(4\omega_0 t) + \ldots.$$

3.2.3 Phasors

Given a periodic signal with fundamental frequency ω_0, we can compute a fundamental frequency phasor from the fundamental term in the Fourier series. If we adopt the notation that a cosine waveform is the reference signal, that is, the voltage

$$v(t) = \sqrt{2}V\cos(\omega_0 t)$$

corresponds to a phasor V which has angle 0, and the voltage

$$v(t) = \sqrt{2}V\cos(\omega_0 t + \varphi)$$

corresponds to a complex phasor, $Ve^{j\phi}$, then the fundamental frequency phasor is directly related to the first exponential Fourier series coefficient:

$$Ve^{j\varphi} = \frac{\sqrt{2}}{T_o} \int_0^{T_o} v(t)e^{-j\omega_0 t}\,dt \tag{3.16}$$

The $\sqrt{2}$ in Equation (3.16) is due to the convention that the magnitude of a phasor is the root-mean-square (rms) value of a sinusoid but can be omitted if a ratio of voltage and current phasors (an impedance calculation) is to be computed.

3.3 Other orthogonal expansions

The property of the exponentials that made it easy to compute the Fourier series coefficients, viz.

$$\int_0^{T_o} e^{j(k-m)\omega_0 t}\,dt = 0;\, k \neq m \tag{3.17}$$

is a reflection of the fact that the exponentials $\{e^{jn\omega_0 t}\}n = 0, \pm 1, \pm 2, \ldots$ are orthogonal over a period $2\pi/\omega_0$. There are other orthogonal periodic functions which have been applied in digital relaying. A family of complex signals $\{\varphi_n(t), n = 1, 2, 3, \ldots\}$ are said to be orthogonal over the interval $0 \leq t \leq T$ if

$$\int_0^T \varphi_n(t)\varphi_m{}^*(t)\,dt = 0;\, m \neq n \tag{3.18}$$

If the signals are real then the complex conjugate is unnecessary in Equation (3.18). The negative sign on m in the exponent of Equation (3.17) is due to the required conjugate.

3.3.1 Walsh functions

The Walsh functions are a set of orthogonal signals on the interval [0,1] which only take on the values ± 1. As such, they seemed appealing for digital implementation since multiplication by a Walsh function involves only algebraic operations. Progress in computer hardware has essentially eliminated the need for Walsh functions.

An expansion of a function $f(t)$ in an arbitrary orthogonal set $\phi_n(t)$, $n = 1, 2, 3, \ldots$ can be thought of as that of finding coefficients such that

$$f(t) = \sum_{n=1}^{N} c_n \varphi_n(t) \tag{3.19}$$

Equation (3.19) is only symbolic since the finite summation will not actually equal $f(t)$ except in exceptional circumstances. However, if Equation (3.19) is multiplied on both sides by $\varphi_m{}^*(t)$ and integrated from 0 to T, the orthogonality of the ϕ_n yields

$$c_m = \frac{\int_0^T \varphi_m{}^*(t) f(t)\, dt}{\int_0^T \varphi_m{}^*(t) \varphi_m(t)\, dt} \tag{3.20}$$

Equation (3.20) is a generalization of Equation (3.9) for signals other than exponentials. For exponentials the denominator of Equation (3.20) equals T_o and

$$\varphi_m{}^*(t) = e^{-jm\omega_o t}$$

3.4   Fourier transforms

The technique of representing a signal as a sum of exponentials can be extended to non-periodic functions through the use of Fourier transforms.[2] The transform pair can be obtained by writing a Fourier series and taking the limit as the period becomes infinite.[3,4] Consider a time limited signal as shown in Figure 3.4, i.e.

$$x(t) = 0; \ |t| \rangle T_1$$

We select a period $T_o \gg T_1$ and let

$$r(t) = \sum_{n=-\infty}^{\infty} x(t - nT_o) \tag{3.21}$$

to form a periodic function made up of shifted replicas of $x(t)$ as shown in Figure 3.5.

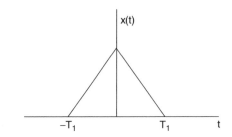

Figure 3.4 A time limited function

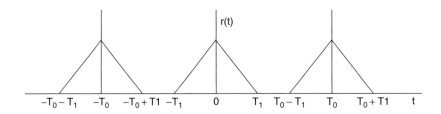

Figure 3.5 Shifted replicas of x(t)

It can be seen that r(t) is periodic, with fundamental period T_0 and fundamental frequency $\omega_o = 2\pi/T_o$. The Fourier series coefficients are given by

$$c_k = \frac{1}{T_o} \int_{-T_0/2}^{T_0/2} r(t)e^{-jk\omega_o t}\,dt$$

$$c_k = \frac{1}{T_o} \int_{-T_0/2}^{T_0/2} x(t)e^{-jk\omega_o t}\,dt$$

As T_o approaches infinity, r(t) limits to x(t), and r(t) can be written

$$r(t) = \sum_{k=-\infty}^{\infty} c_k\, e^{jk\omega_o t}$$

$$r(t) = \sum_{k=-\infty}^{\infty} \left[\frac{1}{T_o} \int_{-T_0/2}^{T_0/2} x(t)e^{-jk\omega_o t}\,dt\right] e^{jk\omega_o t}$$

since

$$\frac{1}{T_o} = \frac{\omega_o}{2\pi}$$

$$r(t) = \frac{1}{2\pi} \sum_{k=-\infty}^{\infty} \left[\int_{-T_0/2}^{T_0/2} x(t)e^{-jk\omega_o t}\,dt\right] \omega_o\, e^{jk\omega_o t}$$

If T_o is taken to infinity in such a way that

$$\omega_o \to d\omega, k\omega_o \to \omega,$$

$$\lim r(t) = x(t),$$

$$T_o \to \infty$$

and $\sum_{k=\infty}^{\infty} \to \int_{-\infty}^{\infty}$

then

$$x(t) = \frac{1}{2\pi} \int_{-\infty}^{\infty} \left[\int_{-\infty}^{\infty} x(t) e^{-j\omega t}\, dt \right] e^{j\omega t}\, d\omega \qquad (3.22)$$

$$x(t) = \frac{1}{2\pi} \int_{-\infty}^{\infty} \hat{X}(\omega) e^{j\omega w t}\, d\omega \qquad (3.23)$$

$$\mathcal{F}\hat{X}(\omega) = \int_{-\infty}^{\infty} x(t) e^{-j\omega t}\, dt \qquad (3.24)$$

$$\mathcal{F}^{-1}x(t) = \frac{1}{2\pi} \int \hat{X}(\omega) e^{j\omega t}\, d\omega \qquad (3.25)$$

Equations (3.24) and (3.25) represent the Fourier transform pair. The connection between the two functions $x(t)$ and $\hat{X}(\omega)$ in Equations (3.24) and (3.25) is shown symbolically as

$$x(t) \overset{\mathcal{F}}{\leftrightarrow} \hat{X}(\omega) \qquad (3.26)$$

Given an $x(t)$ and $\hat{X}(\omega)$, we say $x(t)$ and $\hat{X}(\omega)$ are a Fourier transform pair when either or both of $\mathcal{F}$ (Equation (3.24)) and $\mathcal{F}^{-1}$ (Equation (3.25)) hold.

Example 3.6

If $x(t) = e^{-\alpha t}u(t); \alpha > 0$

$$\hat{X}(\omega) = \int_{-\infty}^{\infty} x(t) e^{j\omega t}\, dt$$

$$= \int_{-\infty}^{\infty} e^{-(\alpha + j\omega)t}\, dt = \frac{1}{\alpha + j\omega}$$

and from the inversion integral

$$x(t) = \frac{1}{2\pi} \int_{-\infty}^{\infty} \frac{e^{j\omega t}}{\alpha + j\omega}\, d\omega$$

Example 3.7

If $x(t) = e^{\alpha t}u(t)$; $\alpha > 0$ then $\hat{X}(\omega)$ does not exist because $x(t) \to \infty$ as $t \to \infty$. It is clear that, since the factor $e^{-j\omega t}$ in the integrand of Equation (3.24) does not contribute to the convergence of the integral, it is necessary that $x(t)$ at least remain bounded as $t \to \infty$ in order for $\mathcal{F}$ & $\mathcal{F}^{-1}$ to be defined.

Example 3.8

If $x(t)$ is the pulse shown in Figure 3.6, i.e.

$$x(t) = p_a(t)$$

$$\hat{X}(\omega) = \int_{-a}^{a} e^{-j\omega t}\, dt = \frac{e^{-j\omega t}}{-j\omega}\Big|_{-a}^{a}$$

$$\hat{X}(\omega) = \frac{2\sin(\omega a)}{\omega}$$

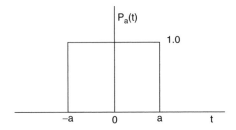

Figure 3.6   Rectangular pulse

In order to plot the transform it is necessary to examine

$$\frac{2\sin(\omega a)}{\omega} \text{ at } \omega = 0$$

Using L'Hospital's rule we evaluate the ratio of the derivatives at $\omega = 0$ to obtain

$$\frac{2\sin(\omega a)}{\omega}\Big|_{\omega=0} = \frac{2a\cos(\omega a)}{1}\Big|_{\omega=0} = 2a$$

The transform $2\sin(\omega a)$ is plotted in Figure 3.7. Using the notation

$$\sin c(x) = \frac{\sin \pi x}{\pi x}$$

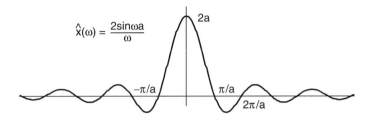

Figure 3.7 The Fourier transform of the rectangular pulse

the transform can be written as

$$\hat{X}(\omega) = 2a \sin c(a\omega/\pi)$$

The example illustrates a general rule. If x(t) is time limited, the Fourier transform will exist. As in the Fourier series case the transform $\hat{X}(\omega)$ is the spectrum of x(t), i.e. the frequency content.

Example 3.9

Suppose $\hat{X}(\omega) = p_{\omega_0}(\omega)$ as shown in Figure 3.8. If it is thought of as the transfer function of a linear system then the filter would represent an 'ideal low-pass filter' since all frequencies below ω_0 are passed without distortion while frequencies above ω_0 are perfectly attenuated. Using the inversion integral (Equation (3.25))

$$x(t) = \frac{1}{2\pi} \int_{-\infty}^{\infty} \hat{X}(\omega)e^{j\omega t}\, d\omega$$

$$x(t) = \frac{1}{2\pi} \int_{-\omega_0}^{\omega_0} e^{j\omega t}\, d\omega$$

$$x(t) = \frac{e^{j\omega t}}{2\pi j t}\bigg|_{-\omega_0}^{\omega_0} = \frac{2\sin(\omega_0 t)}{2\pi t}$$

that is, $X(t) = \dfrac{\sin(\omega_0 t)}{\pi t} = \dfrac{\omega_0}{\pi}\sin c\left(\dfrac{\omega_0 t}{\pi}\right)$

as shown in Figure 3.9. Again if x(t) is band-limited, i.e.

$$\hat{X}(\omega) = 0 \text{ for } |\omega| \rangle \omega_0,$$

then the inverse transform x(t) will always exist.

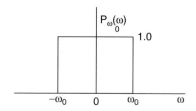

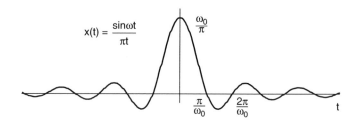

Figure 3.8   The transfer function of the ideal low-pass filter

Figure 3.9   The inverse transform of the ideal low-pass filter shown in Figure 3.8

Example 3.10

If $x(t) = \delta(t)$, the defining properties of the impulse yield

$$\hat{X}(\omega) = \int_{-\infty}^{\infty} \delta(t)e^{-j\omega t} = 1$$

or

$$x(t) = \delta(t) \overset{\mathcal{J}}{\leftrightarrow} 1 = \hat{X}(\omega)$$

The inversion in this pair only makes sense in some generalized context but nonetheless is a transform pair according to our agreements.

Example 3.11

Similarly, if $\hat{X}(\omega) = \delta(\omega)2\pi$

$$F^{-1}x(t) = \frac{1}{2\pi} \int_{-\infty}^{\infty} 2\pi\delta(\omega)e^{j\omega t}\, dt = 1$$

and

$$x(t) = 1 \overset{F}{\leftrightarrow} 2\pi\delta(\omega) = \hat{X}(\omega)$$

We can conclude from the last four pairs of transforms that sharp or narrow signals in one domain (t or ω) imply broad or spread-out signals in the other domain. Sharp edges in time require broad frequency spectra. The examples can also provide a practical view of the impulse. If we normalize Example 3.9 by 2a we obtain

$$P_a(t)/2a \overset{F}{\leftrightarrow} \frac{2\sin(\omega a)}{2a\omega} \tag{3.27}$$

The time function in Equation (3.37) has unit area being 2a wide and 1/2a tall, while the function of frequency has a height at $\omega = 0$ of 1. If a is 10^{-9} sec (roughly the time it takes light to travel a foot) then the pulse has a duration of 2×10^{-9} sec and the transform is as shown in Figure 3.8 with the first zero at $\omega = \pi \times 10^9$ radians or a frequency of 500 M Hz. Practically speaking, the impulse may be regarded as an extremely narrow pulse with unit area (the limit of Equation (3.27) as 'a' goes to zero.)

Example 3.12

If Example 3.10 is shifted in frequency,

$$\hat{X}(\omega) = 2\pi\delta(\omega - \omega_0)$$

$$x(t) = \frac{1}{2\pi} \int_{-\infty}^{\infty} 2\pi\delta(\omega - \omega_0)e^{j\omega t}\, d\omega = e^{j\omega_0 t}$$

$$e^{j\omega_0 t} \overset{F}{\leftrightarrow} 2\pi\delta(\omega - \omega_0)$$

3.4.1 Properties of fourier transforms

There are a number of properties of Fourier transforms which are useful in understanding the resolution of a signal into its frequency spectrum.

P1) Linearity: If $x_1(t) \overset{\mathcal{F}}{\leftrightarrow} \hat{X}_1(\omega)$

And $x_2(t) \overset{\mathcal{F}}{\leftrightarrow} \hat{X}_2(\omega)$

then $c_1 x_1(t) + c_2 x_2(t) \overset{\mathcal{F}}{\leftrightarrow} c_1 \hat{X}_1(\omega) + c_2 \hat{X}_2(\omega)$ $\tag{3.28}$

It should be observed that Equation (3.28) may involve combining one pair where F does not actually exist with a second pair where F^{-1} does not actually exist.

P2) Delay Rule: If $\quad x(t) \overset{F}{\leftrightarrow} \hat{X}(\omega)$

then $\quad x(t - t_1) \overset{F}{\leftrightarrow} \hat{X}(\omega)e^{j\omega t_1}$ $\qquad\qquad\qquad\qquad\qquad$ (3.29)

Equation (3.29) can be verified by evaluating

$$\int_{-\infty}^{\infty} x(t - t_1)e^{j\omega t}\,dt = \int_{-\infty}^{\infty} x(\tau)e^{-j\omega(\tau+t_1)}\,d\tau$$

letting $\quad \tau = t - t_1$

$$= e^{-j\omega t_1}\int_{-\infty}^{\infty} x(\tau)e^{-j\omega\tau}\,dr = \hat{X}(\omega)e^{-j\omega t_1}$$

Example 3.13

If

$$x(t) = P_a(t),\ \hat{X}(\omega) = \frac{2\sin(\omega a)}{\omega}$$

then

$$y(t) = P_a(t - t_1),\ \hat{Y}(\omega) = e^{j\omega t_1}\frac{2\sin(\omega a)}{\omega}$$

If the transform is written $\hat{X}(\omega) = |\hat{X}(\omega)|e^{j\theta(\omega)}$ where $|\hat{X}(\omega)|$ is the magnitude, and $\theta(\omega)$ is the phase, the transforms in Example 3.13, $\hat{X}(\omega)$ and $\hat{Y}(\omega)$, are seen to differ only in phase. Further, the phase of y(t) has a special form

$$\angle\hat{Y}(\omega) = -\omega t_1$$

as shown in Figure 3.10.

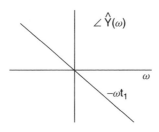

Figure 3.10   Linear phase associated with pure delay

P3) Frequency Shift or Modulation Rule:
If

$$x(t) \overset{\mathcal{F}}{\leftrightarrow} \hat{X}(\omega)$$

then

$$x(t)e^{j\omega_o t} \overset{\mathcal{F}}{\leftrightarrow} \hat{X}(\omega - \omega_o) \tag{3.30}$$

Equation (3.30) can be verified by letting

$$y(t) = x(t)e^{j\omega_o t}$$

$$\hat{Y}(\omega) = \int_{-\infty}^{\infty} x(t)e^{j\omega_o t - j\omega t} \, dt = \int_{-\infty}^{\infty} x(t)e^{-j(\omega - \omega_o)t} \, dt$$

$$\hat{Y}(\omega) = \hat{X}(\omega - \omega_o)$$

Example 3.14

If $y(t) = x(t) \cos(\omega_o t)$ as shown in Figure 3.11, using

$$y(t) = \frac{1}{2}x(t)e^{j\omega_o t} + \frac{1}{2}x(t)e^{-j\omega_o t}$$

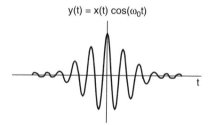

y(t) = x(t) cos(ω₀t)

t

Figure 3.11 An amplitude modulated signal

and the modulation rule

$$\hat{Y}(\omega) = \frac{1}{2}\hat{X}(\omega - \omega_o) + \frac{1}{2}\hat{X}(\omega + \omega_o)$$

The transform pairs are shown in Figure 3.12

P4) Differentiation in Time:
If

$$x(t) \overset{F}{\leftrightarrow} \hat{X}(\omega) \text{ and } \dot{x}(t) \quad \text{exists}$$

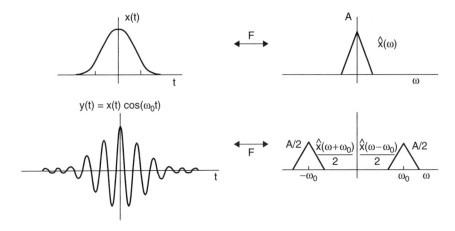

Figure 3.12 Modulation

then

$$\dot{x}(t) \overset{F}{\leftrightarrow} (j\omega)\, \hat{X}(\omega)$$

and

$$\frac{d^n}{dt^n} x(t) \overset{F}{\leftrightarrow} (j\omega)^n\, \hat{X}(\omega) \qquad (3.31)$$

Equation (3.31) is obtained directly by differentiating the inversion integral

$$x(t) = \frac{1}{2\pi} \int_{-\infty}^{\infty} \hat{X}(\omega) e^{j\omega t}\, d\omega$$

$$\frac{dx}{dt} = \frac{1}{2\pi} \int_{-\infty}^{\infty} (j\omega)\, \hat{X}(\omega) e^{j\omega t}\, d\omega$$

P5) Differentiation in Frequency:
If

$$x(t) \overset{\mathcal{F}}{\leftrightarrow} \hat{X}(\omega) \text{ and } \frac{d}{d\omega} \hat{X}(\omega) \text{ exists}$$

then

$$(-jt)x(t) \overset{\mathcal{F}}{\leftrightarrow} \frac{d}{d\omega} \hat{X}(\omega)$$

As in P4:

$$\hat{X}(\omega) = \int_{-\infty}^{\infty} x(t) e^{-j\omega t}\, dt$$

$$\frac{d\hat{X}(\omega)}{d\omega} = \int_{-\infty}^{\infty} (-jt)x(t) e^{-j\omega t}\, dt$$

In general

$$\frac{d^n \hat{X}(\omega)}{d\omega^n} \overset{\mathcal{F}}{\leftrightarrow} (-jt)^n x(t) \tag{3.32}$$

Even and Odd Properties: If x(t) is real

$$\hat{X}(\omega) = \int_{-\infty}^{\infty} x(t) e^{-j\omega t} dt$$

$$\int_{-\infty}^{\infty} x(t) \cos(\omega t) dt - j \int_{-\infty}^{\infty} x(t) \sin(\omega t) dt$$

$$\text{Re}\left\{\hat{X}(\omega)\right\} = \int_{-\infty}^{\infty} x(t) \cos(\omega t) dt : \text{ an even function of } \omega$$

$$\text{Im}\left\{\hat{X}(\omega)\right\} = \int_{-\infty}^{\infty} x(t) \sin(\omega t) dt : \text{ an odd function of } \omega$$

P6a): If x(t) is real and even, i.e. $x(t) = x(-t)$, then $\hat{X}(\omega)$ is real and even.

P6b): If x(t) is real and odd, i.e. $x(t) = -x(-t)$, then $\hat{X}(\omega)$ is pure imaginary and odd.

Properties P6a and P6b follow from

$$\hat{X}(\omega) = \int_0^{\infty} x(t) e^{-j\omega t} dt = \int_{-\infty}^0 x(t) e^{j\omega t} dt$$

$$= \int_0^{\infty} x(t) e^{-j\omega r} dt + \int_0^{\infty} x(-\tau) e^{j\omega \tau} dr$$

$$= \int_0^{\infty} [x(t) e^{-j\omega t} + x(-t) e^{j\omega t}] dt$$

$$= \int_0^{\infty} [x(t) + x(-t)] \cos(\omega t) dt + j \int_0^{\infty} [x(-t) - x(t)] \sin(\omega t) dt$$

$$\text{Re}\{\hat{X}(\omega)\} \qquad\qquad \text{Im}\{\hat{X}(\omega)\}$$

$$= 0 \text{ if } x(t) = -x(-t) \qquad\qquad = 0 \text{ if } x(t) = x(-t)$$

P7) Time Scaling: If $x(t) \overset{\mathcal{F}}{\leftrightarrow} \hat{X}(\omega)$
then

$$x(at) \overset{\mathcal{F}}{\leftrightarrow} \frac{\hat{X}\left(\dfrac{\omega}{a}\right)}{a} ; a \rangle 0 \tag{3.33}$$

To establish Equation (3.33) let y(t) = x(at)

$$\hat{Y}(\omega) = \int_{-\infty}^{\infty} x(at)e^{-j\omega t}\, dt;\ \tau = at;\ t = \tau/a$$

$$dt = \frac{1}{a}\, d\tau$$

$$\hat{Y}(\omega) = \frac{1}{a}\int_{-\infty}^{\infty} x(\tau)e^{-j\left(\frac{\omega}{a}\right)\tau}\, d\tau = \frac{1}{a}\hat{X}\left(\frac{\omega}{a}\right)$$

Equation (3.33) is another expression of the inherent conservation of time-duration and bandwidth. If scaling is used to make the time function *narrow (a>1)* then the transform is scaled in the opposite direction and becomes broader.

Example 3.15

The Gaussian pulse and its transform

$$x(t) = \frac{e^{\frac{-1}{2}\left(\frac{t}{\tau}\right)^2}}{\sqrt{2\pi}\tau} \overset{\mathcal{F}}{\leftrightarrow} \hat{X}(\omega) = e^{\frac{-1}{2}\omega^2\tau^2} \tag{3.34}$$

are shown in Figure 3.13. The transform pair in Equation (3.34) is important in probability and statistics. It further illustrates the time duration-bandwidth limitation, since the width of the time function is proportional to τ while the bandwidth is proportional to $1/\tau$.

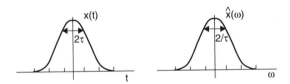

Figure 3.13   The Gaussian pulse and its Fourier transform

P8) Periodic Functions: Using the transforms of the impulse it is possible to discuss the Fourier transforms of periodic functions for which we previously wrote Fourier series. Recall:

$$\delta(t) \overset{\mathcal{F}}{\leftrightarrow} 1$$

$$1 \overset{\mathcal{F}}{\leftrightarrow} 2\pi\, \delta(\omega)$$

$$e^{j\omega_o t} \overset{\mathcal{F}}{\leftrightarrow} 2\pi\, \delta(\omega - \omega_o)$$

$\cos(\omega_0 t)$:

$$\frac{1}{2}[e^{j\omega_0 t} + e^{-j\omega_0 t}] \overset{\mathcal{F}}{\leftrightarrow} \pi[\delta(\omega - \omega_0) + \delta(\omega + \omega_0)]$$

$\sin(\omega_0 t)$:

$$\frac{1}{2j}[e^{j\omega_0 t} - e^{-j\omega_0 t}] \overset{\mathcal{F}}{\leftrightarrow} \frac{\pi}{j}[\delta(\omega - \omega_0) - \delta(\omega + \omega_0)]$$

Fourier series:

$$\sum_{k=-\infty}^{\infty} c_k e^{jk\omega_0 t}; \ \hat{R}(\omega) = 2\pi \sum_{k=-\infty}^{\infty} c_k \ \delta(\omega - k\omega_0) \qquad (3.35)$$

By drawing an impulse of strength c_k with a height of c_k, the transform in Equation (3.35) can be drawn as in Figure 3.14. The discrete spectrum in Figure 3.14 is referred to as a line spectrum. The inversion of Equation (3.25) returns the original expression for the Fourier series:

$$r(t) = \frac{1}{2\pi} \int_{-\infty}^{\infty} 2\pi \sum_{k=-\infty}^{\infty} c_k \delta(\omega - \omega_0) e^{j\omega t} \, d\omega$$

$$r(t) = \sum_{k=-\infty}^{\infty} c_k e^{jk\omega_0 t}$$

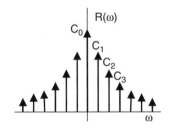

Figure 3.14   A line spectrum

P9): The Fourier transform of the unit step shown in Figure 3.15 must be obtained rather carefully. First let $y(t) = \text{sgn}(t)$ and let $\text{sgn}(t)$ be thought of as the limit shown in Figure 3.16, i.e.

$$\lim_{\alpha \to 0}[e^{-\alpha t} u(t) - u(-t)e^{\alpha t}] = \text{sgn}(t) \qquad (3.36)$$

Using $e^{-\alpha t} u(t) \overset{\mathcal{F}}{\leftrightarrow} \frac{1}{j\omega + \alpha}$

$$e^{-\alpha t} u(-t) \overset{\mathcal{F}}{\leftrightarrow} \frac{1}{\alpha - j\omega}$$

Figure 3.15   The unit step

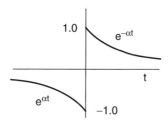

Figure 3.16   A signal which limits to sgn(t)

we obtain

$$\lim_{\alpha \to 0} \left[\frac{1}{\alpha + j\omega} - \frac{1}{\alpha - j\omega} \right] = \frac{2}{j\omega} = \mathcal{F}[\text{sgn}(t)]$$

$$\text{sgn}(t) \overset{\mathcal{F}}{\leftrightarrow} \frac{2}{j\omega} \tag{3.37}$$

It should be noted that the transform pair in Equation (3.37) obeys property P6b in that sgn(t) is odd and the transform 2/jω is pure imaginary.

The transform of the step can be obtained by observing that

$$u(t) = \frac{1}{2} + \frac{1}{2}\text{sgn}(t) \tag{3.38}$$

and hence $u(t) \overset{\mathcal{F}}{\leftrightarrow} \dfrac{1}{j\omega} + \pi\delta(\omega)$ \hfill (3.39)

The $\pi\,\delta(\omega)$ in Equation (3.39) is due to 1/2 in Equation (3.38). In spite of the fact that the Laplace transform of the unit step is 1/s the Fourier transform contains the $\pi\,\delta(\omega)$ term in addition to 1/ω.

P10) Convolution in Time:
Given two signals with Fourier transforms

$$x_1(t) \overset{\mathcal{F}}{\leftrightarrow} \hat{X}_1(\omega), \, x_2(t) \overset{\mathcal{F}}{\leftrightarrow} \hat{X}_2(\omega)$$

and a well defined convolution $x_1 * x_2$ then

$$x_1 * x_2 \overset{\mathcal{F}}{\leftrightarrow} \hat{X}_1(\omega)\hat{X}_2(\omega) \qquad (3.40)$$

That is, convolution in time corresponds to multiplication in the transform domain. Equation (3.40) can be verified by taking the transform of the convolution. Let

$$y(t) = x_1 * x_2$$

$$y(t) = \int_{-\infty}^{\infty} x_1(\tau)x_2(t - \tau)\, d\tau$$

then

$$\hat{Y}(\omega) = \int_{-\infty}^{\infty} \left[\int_{-\infty}^{\infty} x_1(\tau)x_2(t - \tau)\, d\tau \right] e^{-j\omega t}\, dt$$

$$\hat{Y}(\omega) = \int_{-\infty}^{\infty} x_1(\tau)\hat{X}_2(\omega)e^{-j\omega t}\, d\tau \text{ (by delay rule P2)}$$

$$\hat{Y}(\omega) = \hat{X}_1(\omega)\hat{X}_2(\omega)$$

Two examples of convolution given below are particularly interesting.

Example 3.16
Let

$$Y(t) = x(t) * u(t) = \int_{-\infty}^{t} x(\tau)\, dr$$

$$\hat{Y}(\omega) = \hat{X}(\omega)\left[\pi\delta(\omega) + \frac{1}{j\omega} \right]$$

$$\hat{Y}(\omega) = \frac{\hat{X}(\omega)}{j\omega} + \pi\delta(\omega)\,\hat{X}(0)$$

Example 3.17
If $x(t) * x(-t)$ is defined
then

$$y(t) = x(t) * x(-t) = \int_{-\infty}^{\infty} x(\tau)x(-t + \tau)\, d\tau$$

i.e.

$$\hat{Y}(\omega) = \hat{X}(\omega)\hat{X}^*(\omega) = |\hat{X}(\omega)^2|$$

However, if x(t) is real

$$F[x(-t)] = \int_{-\infty}^{\infty} x(-t)e^{-j\omega t}\,dt = \int_{-\infty}^{\infty} x(\tau)e^{j\omega\tau}\,dr = \hat{X}^*(\omega)$$

$$y(t) = \frac{1}{2\pi}\int_{-\infty}^{\infty} |\hat{X}(\omega)|^2 e^{j\omega t}\,dt$$

$$y(0) = \int_{-\infty}^{\infty} x^2(\tau)\,d\tau = \frac{1}{2\pi}\int_{-\infty}^{\infty} |\hat{X}(\omega)|^2\,d\omega$$

which is Parseval's theorem for non-periodic functions.

P11) Convolution in Frequency:
Given

$$x_1(t) \overset{\mathfrak{F}}{\leftrightarrow} \hat{X}_1(\omega), \quad x_2(t) \overset{\mathfrak{F}}{\leftrightarrow} \hat{X}_2(\omega)$$

with $\hat{X}_1(\omega) * \hat{X}_2(\omega)$ defined,
then

$$x_1(t)x_2(t) \overset{\mathfrak{F}}{\leftrightarrow} \frac{1}{2\pi}\hat{X}_1(\omega) * \hat{X}_2(\omega) \tag{3.41}$$

Equation (3.41) is dual to Equation (3.40) in that convolution in one domain corresponds to multiplication in the other domain. Equation (3.41) can be verified by taking the inverse transform of the convolution in frequency defined as

$$\hat{Y}(\omega) = \frac{1}{2\pi}\int_{-\infty}^{\infty} \hat{X}_1(\mu)\hat{X}_2(\omega - \mu)\,d\mu$$

and using the modulation rule, P3, where the delay rule, P2, was used in establishing Equation (3.40).

Example 3.18
The ideal low-pass filter from Example 3.9

$$P_a(\omega) \overset{F}{\leftrightarrow} \frac{\sin(at)}{\pi t}$$

generates the pair

$$\frac{\sin^2(at)}{\pi^2 t^2} \overset{F}{\leftrightarrow} \frac{1}{2\pi} P_a(\omega) * P_a(\omega)$$

using P11. The transform pair is shown in Figure 3.17. It can be seen that multiplying time functions increases the bandwidth since the convolution of two band-limited spectra with bandwidths ω_1 and ω_2 respectively produces a transform with bandwidth $\omega_1 + \omega_2$.

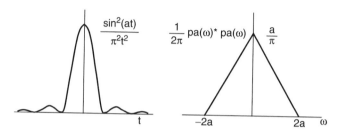

Figure 3.17   The transform pair from Example 3.18

P12) Symmetry or Duality:

The symmetry property allows us to effectively double the number of transform pairs at our disposal, in that, for each pair we know, we can find a second pair in the following manner:

If

$$f(t) \overset{\mathcal{F}}{\leftrightarrow} g(\omega) \tag{3.42}$$

then

$$g(t) \overset{\mathcal{F}}{\leftrightarrow} 2\pi f(-\omega) \tag{3.43}$$

The notation used in Equations (3.42) and (3.43) is slightly different from that used previously in order to avoid confusion between the two domains. Here, for example, the function g(.) is a transform in Equation (3.42) while it is a time function in Equation (3.43).

The property can be established through a sequence of changes of variables in the transform equation:

$$g(\omega) = \int_{-\infty}^{\infty} f(t)e^{-j\omega t}\, dt$$

or

$$g(z) = \int_{-\infty}^{\infty} f(p)e^{-jzp}\, dp$$

Now let $-p = \omega$, $z = t$

$$g(t) = \int_{\infty}^{-\infty} f(-\omega)e^{jt\omega}(-d\omega)$$

$$g(t) = \frac{1}{2\pi} \int_{-\infty}^{\infty} 2\pi f(-\omega)e^{j\omega t}\, d\omega$$

Example 3.19

Given $x(t) = \dfrac{1}{1+t^2}$

Using duality if we consider

$$g(\omega) = \frac{1}{1+\omega^2} = \frac{1}{2}\left[\frac{1}{1+j\omega} + \frac{1}{1-j\omega}\right]$$

i.e. $\frac{1}{1+\omega^2} \overset{F}{\leftrightarrow} \frac{1}{2}e^{-|t|}$ then from symmetry

$$g(t) = \frac{1}{1+t^2} \overset{F}{\leftrightarrow} 2\pi[f(-\omega)] = \pi e^{-|\omega|}$$

3.5 Use of fourier transforms

Fourier transforms are used in a variety of areas ranging from communication systems to optics. The immediate application of interest in relaying is in the description of the frequency content of signals and of the effect of filters and algorithms on those signals. If the box in Figure 3.18 represents a Linear Time Invariant System (LTI) then there is a function h(t) such that

$$y(t) = h(t) * w(t) \tag{3.44}$$

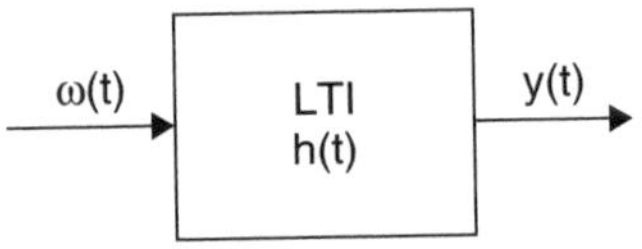

Figure 3.18 LTI system

If the input is of the form $w(t) = e^{j\omega t}$ then, using ideas from elementary circuit analysis, the output is of the same form, i.e.

$$y(t) = \hat{H}(\omega)e^{j\omega t} \tag{3.45}$$

Taking the Fourier transform of Equation (3.44), it is also clear that

$$\hat{Y}(\omega) = \hat{H}(\omega)\,\hat{W}(\omega) \tag{3.46}$$

where

impulse response $h(t) \overset{\mathcal{F}}{\leftrightarrow} \hat{H}(\omega)$ frequency response.

In other words, the frequency response of the system determines how the frequency content of the input is passed to the output. It is in this sense that the function of frequency of Example 3.9 can be considered to be an ideal low-pass filter. The filters used in relaying applications have been discussed in Chapter 1. We will use the ideal low-pass filter for convenience in discussing the idea of sampling.

3.5.1 Sampling

Given that $x(t)$ is a band-limited signal, i.e.

$$x(t) \overset{\mathcal{F}}{\leftrightarrow} \hat{X}(\omega)$$

and

$$\hat{X}(\omega) = 0 \text{ for } |\omega| > \omega_m$$

and suppose $x(t)$ were multiplied by the periodic signal shown in Figure 3.19 to produce a new signal called the sampled signal,

$$r(t) = \sum_{k=-\infty}^{k=\infty} p_a(t - nT_o) \text{ with } T_o - a > a, \text{ or } T_o > 2a$$

If the product of $x(t)$ and $r(t)$ is formed

$$z(t) = x(t)r(t)$$

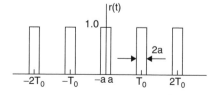

Figure 3.19 Sampling signal

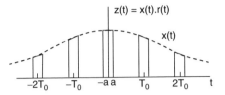

Figure 3.20   Sampled signal

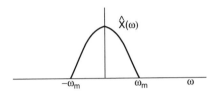

Figure 3.21   A band-limited spectrum

as shown in Figure 3.20 it is possible to recover the signal $x(t)$ from the signal $z(t)$
with a filter if the samples are close enough together.

The process is best understood in the frequency domain. If the spectrum of $x(t)$
is as shown in Figure 3.21 and we form the transform of $z(t)$ using the convolution
in frequency rule, we obtain

$$\hat{Z}(\omega) = \frac{\hat{X}(\omega) * \hat{R}(\omega)}{2\pi}$$

Since $r(t)$ is a periodic function, it has a line spectrum, i.e. with $\omega_o = 2\pi/T_o$

$$r(t) = \sum_{k=-\infty}^{k=\infty} c_k e^{jk\omega_o t} \overset{F}{\leftrightarrow} 2\pi \sum_{k=-\infty}^{k=\infty} c_k \delta(\omega - k\omega_o)$$

and

$$\hat{Z}(\omega) = \frac{2\pi}{2\pi} \sum_{k=-\infty}^{k=\infty} c_k \hat{X}(\omega) * \delta(\omega - k\omega_o)$$

if we examine the convolution with impulses

$$\hat{X}(\omega) * \delta(\omega - k\omega_o) = \int_{-\infty}^{\infty} \delta(\mu - k\omega_o) \hat{X}(\omega - \mu) \, d\mu$$

$$= \hat{X}(\omega) - k\omega_o)$$

$$\hat{Z}(\omega) = \sum_{k=-\infty}^{k=\infty} c_k \hat{X}(\omega - k\omega_o) \tag{3.47}$$

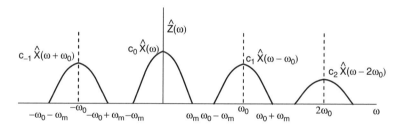

Figure 3.22 The spectrum of the sampled signal

Equation (3.47) represents a spectrum made up of shifted replicas of $\hat{X}(\omega)$ as shown in Figure 3.22. If, as shown,

$$\omega_m < \omega_0 - \omega_m$$

i.e.

$$\omega_0 > 2\omega_m$$

or

$$T_0 < \pi/\omega_m$$

then the spectrum of x(t) can be recovered from z(t) with an ideal low-pass filter with a cut-off frequency of ω_m.

The frequency, $\omega_0 = 2\omega_m$, i.e. twice the highest frequency in the band-limited signal, is the Nyquist sampling frequency. It can be seen from Figure 3.22 that if the sampling frequency were lower than the Nyquist frequency that there would be overlap in the shifted replicas of $X(\omega)$ and the output of the low-pass filter would not be the original signal x(t). This effect is called aliasing. If a signal is to be sampled at a sampling rate corresponding to the frequency ω_s then to avoid aliasing it is necessary to filter the signal to a bandwidth of $\omega_s/2$. Such a filter is referred to as an anti-aliasing filter (as discussed in Chapter 1).

3.6 Discrete fourier transform

Given a continuous signal f(t), if we take N samples of f(t) at intervals of T sec to form a finite duration discrete time signal as shown in Figure 3.23 with

$$f[n] = f(nT) \quad 0 \langle n \langle N - 1$$

$$\hat{F}(\omega) = \sum_{n=0}^{N-1} d(nT)e^{-j\omega T} \tag{3.48}$$

then the transform, $\hat{F}(\omega)$, is a continuous and periodic function of ω with a Fourier series with only a finite number of terms (N) as shown by Equation (3.48).[5] The time function is described by N numbers, i.e.

$$f[0], f[1], f[2], \ldots, f[N-1]$$

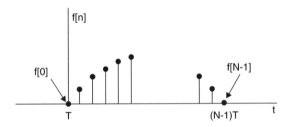

Figure 3.23   Finite duration discrete time signal

while the function of ω is continuous. It would be desirable (if possible) to obtain a finite description of the transform, i.e. N numbers which completely describe the transform, $\hat{F}(\Omega)$. The key is to consider the periodic extension (see Problem 3.3) of the time-limited discrete time signal. It can be shown that the discrete Fourier transform pair

$$\text{DFT}: \qquad \hat{F}(k\Omega_o) = \sum_{n=0}^{N-1} f(nT)e^{-jkn\Omega_o T} \qquad (3.49)$$

$$\text{IDFT}: \qquad f(nT) = \frac{1}{N} \sum_{k=0}^{N-1} \hat{F}(k\Omega_o)e^{jnk\Omega_o T} \qquad (3.50)$$

relate a sampled periodic time function (the periodic extension of the signal in Figure 3.23) and a similar sampled periodic function of ω as shown in Figure 3.24. There are N samples per period in each domain. In the time domain the samples are at an interval of T sec and the period is NT sec. In the frequency domain the samples are spaced at $2\pi/NT = \Omega_0$ radians and the period is $2\pi/T$.

A more compact form of Equations (3.49) and (3.50) can be obtained if we let

$$F_k = \hat{F}(k\Omega_o) \qquad f_n = f(nT)$$

$$w = e^{-j\Omega_o T} = e^{-j\frac{2n}{N}}$$

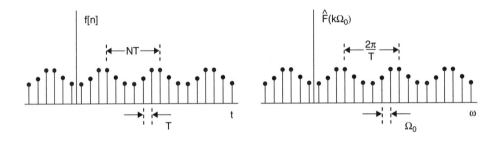

Figure 3.24   Discrete Fourier transform pair

then

$$f_k = \sum_{n=0}^{N-1} f_n w^{nk} \qquad (3.51)$$

$$f_n = \frac{1}{N} \sum_{k=0}^{N-1} F_k w^{-nk} \qquad (3.52)$$

which makes the connection between N numbers in time and N numbers in frequency more obvious. Notice that the difference in sign of the exponents plays the same role as in the original transform and that the inverse transform has a scale factor of $1/N$ rather than $1/2\pi$. Equations (3.51) and (3.52) can also be expressed in matrix form as

$$[EQ] \begin{bmatrix} F_o \\ F_1 \\ \vdots \\ \vdots \\ F_{N-1} \end{bmatrix} = \begin{bmatrix} w^0 & w^0 & \cdots\cdots & w^0 \\ w^0 & w^1 & w^2 & \cdots \\ w^0 & w^2 & w^4 & \cdots \\ w^0 & w^3 & w^6 & \cdots \end{bmatrix} \begin{bmatrix} f_o \\ f_1 \\ \vdots \\ f_{N-1} \end{bmatrix} \qquad (3.53)$$

In applications other than digital relaying, the fast Fourier transform (FFT) has become the vehicle of application of the transform techniques developed in the first part of this chapter. The FFT is in fact not a new transform, but rather a numerical technique to make the calculation of the DFT faster.[6] It is orders of magnitude faster than the calculations implied by (Equation (3.53)) if all the F_k are desired and if N is large. In relaying applications, however, N is small (from 4 to 20 for most algorithms) and only a few of the F_k are wanted. Generally only the fundamental frequency component ($k = 1$) is used in impedance relaying while a few harmonics (for example, the fundamental, the second, and the fifth) are used in transformer algorithms. Hence the FFT has found little application in digital relaying.

If we consider the signal

$$v(t) = \sqrt{2}V \cos(\omega t + \phi)$$

with N samples per period ($T = 2\pi/\omega$), the calculation in Equation (3.49) yields

$$F_1 = \frac{N/2}{2} V e^{j\varphi}$$

The fundamental frequency phasor associated with the signal x(t) is then given by

$$X_1 = \frac{2}{2/N} \sum_{n=0}^{n=N-1} x\left(\frac{2\pi n}{N\omega}\right) e^{-j2\pi n/N}$$

or with

$$x_n = x\left(\frac{2\pi n}{N\omega}\right)$$

$$X_1 = \frac{2}{2/N}\sum_{n=0}^{n=N-1} x_n e^{-j2\pi n/N} \qquad (3.54)$$

Equation (3.54) provides the basis of a number of relaying algorithms and will be seen again in Chapter 4.

3.7 Introduction to probability and random process

As pointed out in Chapter 1, errors in the analog to digital conversion process can be thought of as being random. There are additional errors or unmodeled signals present in most relaying applications as will be discussed in later chapters. These include signals not described by the model on which the algorithm is based such as: the transient response of the capacitively coupled voltage transformers, harmonics produced by CT saturation, high frequency signals associated with reflection of surge waveforms between the bus and the fault, and other system transients caused by the fault. One valid comparison of algorithms is in terms of their ability to reject these unwanted signals. If these error signals are considered to be random processes then techniques of parameter estimation can be used to describe and evaluate algorithms. This section is concerned with the probability and random processes underlying this view of relaying algorithms.[7]

3.7.1 Random variables and probability distributions

Consider an experiment in which a die is thrown and the result recorded. Suppose the experiment is repeated 600 times. A typical set of results is summarized in Table 3.1.

Table 3.1 Results of experiment

Outcome	Number
1	101
2	100
3	95
4	94
5	113
6	97

If we define the relative frequency of the outcome being a 3, for example, as

$$F3 = \frac{\text{number of times a 3 is observed}}{\text{Total number of throws}}$$

$$F3 = 95/600$$

then, as the number of throws, N, increases, a belief in statistical regularity would imply that F_3 should limit to a constant. We could define the probability that a 3 is observed as that limit, i.e.

$$\text{Pr}\{a\ 3\ \text{is observed}\} = \lim F_3$$

For a 'fair' die we would expect all of the probabilities to be 1/6. The outcome of the experiment is then a random variable with six possible values 1, 2, 3, 4, 5, and 6, and we have assigned a probability to each of the six possible outcomes.

If we imagine a similar but more elaborate experiment we can introduce a more compact description of the behavior of the random variable. Suppose we throw 15 dice (a real handful) and record the total of the faces showing. The result can vary from 15 (all 1's) to 90 (all 6's). If we attempt to summarize the result in a table such as Table 3.1 we will need 76 entries (the 76 possible outcomes). A more compact summary can be given by making 76 bins and counting the number of times the total lies in each bin. If we present the results in a histogram we get a curve such as that in Figure 3.25 (drawn for 30 000 trials). Another possible way of displaying the result is to count the number of outcomes with values less than or equal to each value. Such a cumulative histogram is shown in Figure 3.26.

3.7.2   *Probability distributions and densities*

In general, random variables can be described with functions similar to the cumulative histogram of Figure 3.26. If X denotes the random variable then the

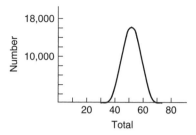

Figure 3.25   Histogram

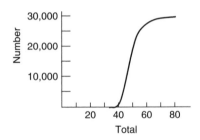

Figure 3.26 Cumulative histogram

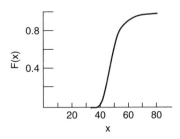

Figure 3.27 A probability distribution function

function $F_X(x)$ defined as

$$F_X(x) = \Pr\{X \le x\} \tag{3.55}$$

is called the probability distribution function for X. It is clear that $F_x(x)$ is a monotone non-decreasing function of x. The corresponding distribution function for our previous experiment is shown in Figure 3.27.

If a function $f(x)$ exists such that

$$F_X(x) = \int_{-\infty}^{x} f(\xi)\, d\xi \tag{3.56}$$

or

$$f(x) = \frac{dF(x)}{dx}$$

then $f(x)$ is referred to as the probability density function for X. It is the existence of probability distribution functions and density functions which are tractable for analysis that has made most of the results in modern probability and statistics possible. Two common densities are the Gaussian density and the uniform density. The Gaussian or normal density is given by

$$f(x) = \frac{1}{\sqrt{2\pi}\sigma} e^{-(x-m)^2/2\sigma^2} \tag{3.57}$$

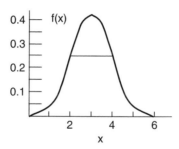

Figure 3.28   Gaussian density

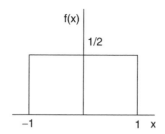

Figure 3.29   Uniform density

and is shown in Figure 3.28 for an m of 3.0 and a σ of 1.0. A uniform density is shown in Figure 3.29.

3.7.3 Expectation

Given a function, g(x), of the random variable X, the expectation of g(x) is defined as

$$E\{g(x)\} = \int_{-\infty}^{\infty} g(x)f(x)\,dx \tag{3.58}$$

In particular, the expected value of powers of x are important in describing the random variable. The expected value of x is called the mean and is given by

$$\bar{x} = E\{x\} = \int_{-\infty}^{\infty} xf(x)\,dx \tag{3.59}$$

For the Gaussian density, $\bar{x} = m$ (m = 3 in Figure 3.28). The fact that the mean of the uniform distribution in Figure 3.29 is 0 can be verified by integration.

The expected value of the square of the difference between x and its mean is a measure of the width of the density. The variance of the random variable x is defined as

$$\sigma^2 = E\{x - \bar{x})^2\} = \int_{-\infty}^{\infty} (x - \bar{x})^2 f(x)\, dx \qquad (3.60)$$

The quantity σ is referred to as the standard deviation. The standard deviation for the Gaussian density is, of course, σ. The standard deviation of the Gaussian density in Figure 3.28 is 1. The width of the density at the one-σ level is shown in Figure 3.28. Note that

$$F(m + \sigma) = e^{-1/2} f(m)$$

$$e^{-1/2} = .6075$$

The variance of the uniform density in Figure 3.30 can be obtained as

$$\sigma^2 = \int_{-1}^{1} \frac{1}{2} x^2\, dx = \frac{1}{2} \frac{x^3}{3}\Big|_{1}^{-1} = \frac{1}{3}$$

or the standard deviation *is* $\sigma = .57735$.

3.7.4 *Jointly distributed random variables*

It is common to have more than one source of random error in a given application. For that reason we must be concerned with jointly distributed random variables. Because of our intended use of these results we will step directly to a vector valued random variable. Let **X** be a vector of n random variables

$$\mathbf{X} = \begin{vmatrix} X_1 \\ X_2 \\ \vdots \\ X_n \end{vmatrix}$$

with a joint probability distribution

$$F_X(x) = Pr\{X \le x\}$$

where x is also an n vector. There is also a joint density of the form

$$f(x) = \frac{\partial^n F(x)}{\partial x_1 \partial x_2 \ldots \partial x_n}$$

Expectations are computed with repeated integrals in the form

$$E\{g(x)\} = \int\limits_{-\infty}^{\infty} \int\limits_{-\infty}^{\infty} \cdots \int\limits_{-\infty}^{\infty} g(x)f(x) \; dx_1 \; dx_2 \ldots \; dx_n$$

In particular, the mean $\bar{x}$ is a vector and the covariance matrix P is defined as

$$P = E\{(x - \bar{x})(x - \bar{x})^T\} \tag{3.61}$$

where the superscript T denotes the transpose of the vector, i.e. P is a symmetric n by n matrix with

$$P_{ij} = E\{(x_i - \bar{x}_i)(x_j - \bar{x}_i)\}$$

The diagonal entries in P are the variances of the individual random variables, while the off-diagonal entries are in some sense a measure of the connection between the random variables. The form of the density for a Gaussian random variable is

$$F(x) = \frac{\exp\left[-\frac{1}{2}(x - m)^T P^{-1}(x - m)\right]}{(2\pi)^{n/2} \det P^{1/2}} \tag{3.62}$$

Example 3.20

Given two Gaussian random variables with $E\{X_1\} = 0$, $E\{X_2\} = 0$, $E\{X_1^2\} = 5$, $E\{X_2^2\} = 3$, and $E\{X_1 X_2\} = -3$,
then

$$P = \begin{bmatrix} 5 & -3 \\ -3 & 5 \end{bmatrix}, \quad P^{-1} = \frac{1}{6}\begin{bmatrix} 3 & 3 \\ 3 & 5 \end{bmatrix}$$

and

$$f(x_1, x_2) = \frac{1}{\sqrt{24\pi}} \exp\left[-\frac{1}{12}(3x_1^2 + 6x_1 x_2 + 5x_2^2)\right]$$

3.7.5 Independence

Random variables are said to be independent if their densities or distributions factor into a product of densities of distributions. For example X_1 and X_2 are independent if

$$f_{X_1 X_2}(x_1, x_2) = f_{X_1}(x_1) f_{X_2}(x_2)$$

Since the joint Gaussian density in Equation (3.62) is completely specified by its mean and covariance matrix, it follows that Gaussian random variables are independent if their covariance matrix is diagonal. This is not true of arbitrary densities but is one of the many conveniences of the Gaussian density.

3.7.6 Linear estimation

As we will see in Chapter 4, many algorithms involve processing a total number of measurements that exceeds the number of parameters to be determined. In its simplest form such a problem can be cast as that of solving an 'overdefined' set of equations in the form

$$A x = b \tag{3.63}$$

where A and b are known and x is to be determined. The equations are 'overdefined' if there are more b's than x's. As an example consider

$$\begin{bmatrix} 1 & 0 \\ 1 & -1 \\ 0 & 1 \end{bmatrix} \begin{bmatrix} x_1 \\ x_2 \end{bmatrix} '' = '' \begin{bmatrix} 3/4 \\ -1/4 \\ 3/4 \end{bmatrix}$$

The equal sign is in quotes since there is no solution to the equation. The first row has solution $x_1 = 5/4$ while the last row gives $x_2 = 3/4$. Under these conditions the second row is

$$1/2 = -1/4$$

A more reasonable approach to the 'solution' of an equation such as Equation (3.63) is to recognize that there is an error and write

$$b = A x + e$$

where

$$e = b - Ax$$

The solution above is one in which e(1) and e(3) are 0 and e(2) is $-3/4$. In an attempt to spread the error around a little more we could take as a measure of the quality of the solution the sum of the squared e's, i.e.

$$\begin{aligned} e^T e &= (b - Ax)^T (b - Ax) \\ &= (b^T - x^T A^T)(b - A) \\ &= x^T A^T A\, x - x^T A^T b - b^T A\, x + b^T b \end{aligned} \tag{3.64}$$

The x that minimizes $e^T e$ can be obtained by taking the partial derivatives of Equation (3.74) with respect to the components of x and equating to zero. The result is

$$\hat{X} = (A^T A)^{-1} A^T b \tag{3.65}$$

The calculation in Equation (3.65) is sometimes referred to as the pseudo inverse. A mnemonic for Equation (3.65) is to multiply Equation (3.63) on both sides by A^T and then multiply by the inverse of the square matrix $(A^T A)$, i.e.

$$A\hat{x} = b$$
$$A^T A\hat{x} = A^T b$$
$$\hat{x} = (A^T A)^{-1} A^T b$$

For our example

$$A^T A = \begin{bmatrix} 1 & 1 & 0 \\ 0 & -1 & 1 \end{bmatrix} \begin{bmatrix} 1 & 0 \\ 1 & -1 \\ 0 & 1 \end{bmatrix} = \begin{bmatrix} 2 & -1 \\ -1 & 2 \end{bmatrix}$$

$$(A^T A)^{-1} = \frac{1}{3} \begin{bmatrix} 2 & 1 \\ 1 & 2 \end{bmatrix}$$

$$A^T b = \begin{bmatrix} 1 & 1 & 0 \\ 0 & -1 & 1 \end{bmatrix} \begin{bmatrix} 5/4 \\ -1/4 \end{bmatrix} = \begin{bmatrix} 1 \\ 1 \end{bmatrix}$$

$$\hat{x} = \begin{bmatrix} 1 \\ 1 \end{bmatrix}, \quad b - A\hat{x} = \begin{bmatrix} 1/4 \\ -1/4 \\ -1/4 \end{bmatrix}$$

It is an accident that the errors are of exactly the same magnitude in the example, but it is reasonable to assume that they will be approximately the same size since we are minimizing the sum of the squares of all of the errors.

3.7.7 Weighted least squares

Suppose we had a more detailed statistical knowledge of the errors in the equation

$$b = A x + e$$

Suppose we knew that e had a zero mean and a covariance matrix

$$E\{e\, e^T\} = V \tag{3.66}$$

If e were Gaussian and the covariance matrix, V, were diagonal, the components of the error would be independent (but possibly of different sizes). Given a covariance, V, it would make more sense to weight the errors in the minimization, i.e. to seek an x that minimized

$$e^T V^{-1} e$$

Again, if V were diagonal

$$e^T V^{-1} e = \sum \frac{e^2(i)}{V_{ii}}$$

That is, an error e(i) with a large variance would have a smaller contribution to the sum than one with a smaller covariance. The solution to the minimization is given by

$$\hat{X} = (A^T V^{-1} A)^{-1} A^T V^{-1} b \tag{3.67}$$

For the example, if

$$V = \begin{vmatrix} 1 & 0 & 0 \\ 0 & 1 & 0 \\ 0 & 0 & 100 \end{vmatrix}$$

$$\hat{x} = \begin{vmatrix} 1.2426 \\ 1.4853 \end{vmatrix}, \quad b - A\hat{x} = \begin{vmatrix} .0074 \\ -.0074 \\ -.7353 \end{vmatrix}$$

It can be seen that, since the third error is assumed to have a variance 100 times the other two errors, the estimate x̂ drives the error in the first two components down to much smaller values than the third. In fact, the first two errors are much smaller than they were when the three errors were treated equally.

Weighted least squares estimation (Equation (3.67)) is used in power system static state estimation and is an ingredient of many relaying algorithms.

3.8   Random processes

Consider a time function that depends on one or more random variables. In Chapter 4 we will see that the non-fundamental frequency component of the fault waveform can be modeled as depending on random parameters such as the fault incidence angle and fault location. Suppose we imagine an ensemble of simultaneous experiments where the random variables are chosen according to their probability laws as represented in Figure 3.30. Each time function is a sample function of the random process. If all of the sample functions are examined at a fixed time t_i, as shown, then the random variable $x(t_i)$ can be described by a probability density.

The joint probability density for x at all times t_i is then a description of the random process. Somewhat more convenient characterizations of the process are possible if the process is well behaved. For example the mean of the process can be computed

$$\hat{x}(t) = E\{x(t)\}$$

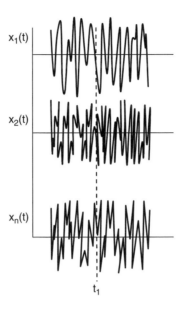

Figure 3.30   An ensemble of experiments

and a correlation function defined

$$R_x(t_1, t_2) = E\{x(t_1)x(t_2)\} \tag{3.68}$$

The process is said to be stationary in the wide sense if the mean is constant and the correlation function depends only on the difference between t_1 and t_2, i.e.

$$R_x(t, t - \tau) = R_x(\tau) \tag{3.69}$$

The Fourier transform of the correlation function of a wide sense stationary random process has a special meaning. Let the power spectrum of the random process $x(t)$ be defined as

$$S_x(\omega) = \int_{-\infty}^{\infty} R_x(\tau)e^{-j\omega t}\, d\tau \tag{3.70}$$

$$R_x(\tau) = \frac{1}{2\pi} \int_{-\infty}^{\infty} S_x(\omega)e^{j\omega \tau}\, d\omega \tag{3.71}$$

In other words, the correlation function and the power spectrum are a Fourier transform pair. The power spectral density is a description of how the energy in the random process is distributed in frequency. A process that has a power spectral density that is a constant for all ω is referred to as 'white noise' since it contains

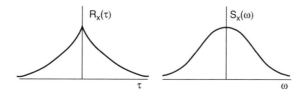

Figure 3.31　A correlation function and power density function

all frequencies. A correlation function and the corresponding power spectrum are shown in Figure 3.31.

Example 3.21

Consider the random process given by

$$x(t) = A \cos(\omega\tau + \phi)$$

where A is a constant, ω is a uniform random variable in the interval from ω_1 to ω_2, and φ is independent of ω and uniform in 0 to 2π. This is similar to a model of the transient components of voltages and currents in a power network. Each sample function is a pure cosine with a constant frequency and amplitude. The expectation in the calculation of the correlation function in Equation (3.69) is over both ω and φ. That is,

$$R_x(t_1, t_2) = \frac{1}{2\pi(\omega_2 - \omega_1)} \int_0^{2\pi} \int_{\omega_1}^{\omega_2} A^2 \cos(\omega t_1 + \varphi) \cos(\omega t_2 + \varphi) \, d\omega \, d\varphi$$

Using

$$\cos x \cos y = \frac{1}{2}\cos(x + y) + \frac{1}{2}\cos(x - y)$$

$$R_x(t_1, t_2) = \frac{A^2}{4\pi(\omega_2 - \omega_1)} \int_0^{2\pi} \int_{\omega_1}^{\omega_2} \cos(\omega(t_2 + t_1) + 2\varphi) \, d\varphi \, d\omega$$

$$+ \frac{A^2}{4\pi(\omega_2 - \omega_1)} \int_0^{2\pi} \int_{\omega_1}^{\omega_2} \cos(\omega(t_2 - t_1)) \, d\omega \, d\varphi$$

The first integral is zero and the second is a function of the difference, $\tau = t_2 - t_1$.

$$R_x(\tau) = \frac{A^2}{2(\omega_2 - \omega_1)} \left[\frac{\sin \omega_2 \tau}{\tau} - \frac{\sin \omega_1 \tau}{\tau} \right]$$

Using the transform of the ideal low-pass

$$S_x(\omega) = \frac{\pi A^2}{(\omega_2 - \omega_1)}; \quad -\omega_2 \le \omega \le -\omega_1,\, \omega_1 \le \omega \le \omega_2$$

$$= 0; \text{ elsewhere.}$$

The spectrum is shown in Figure 3.32.

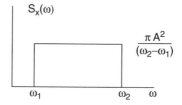

Figure 3.32   The power density spectrum of Example 3.20

3.8.1   Filtering of random processes

If a wide sense stationary random process $x(t)$ with power density spectrum $S_x(\omega)$ is the input to a filter with transfer function $H(\omega)$ as shown in Figure 3.33 then the output $y(t)$ is also a wide sense stationary random process with

$$S_y(\omega) = S_x(\omega)\,|H(\omega)|^2 \tag{3.72}$$

Equation (3.72) can be established by writing

$$y(t) = \int_{-\infty}^{\infty} x(\tau) h(t - \tau)\, d\tau$$

forming $E\{y(t)y(t+\tau)\}$, and using Fourier transforms. Our particular interest in the relationship in Equation (3.72) is in terms of the anti-aliasing filter. If $H(\omega)$ is the transfer function of the anti-aliasing filter and $x(t)$ is the raw (unfiltered) measurement noise then $y(t)$ is the noise seen by the A/D converter. The multiplication in Equation (3.72) then alters the power density spectrum of the noise. For example,

Figure 3.33   Filtering of the random process $x(t)$

if $x(t)$ has the spectrum shown in Figure 3.32 with $\omega_1 = 0$ and the filter $H(\omega)$ is an ideal low-pass with a cut-off frequency of ω_3 ($\omega_3 < \omega_2$) then the output power spectrum is also flat but has a cut-off frequency of ω_3 radians. The impact of anti-aliasing filters on the performance of line relay algorithms will be investigated in more detail in Chapter 4.

3.9 Kalman filtering

The Kalman filter provides a solution to the estimation problem in the context of an evolution of the parameters to be estimated according to a state equation. It has been used extensively in estimation problems for dynamic systems.[8–10] Its use in relaying is motivated by the filter's ability to handle measurements that change in time. To model the problem so that a Kalman filter may be used it is necessary to write a state equation for the parameters to be estimated in the form

$$x_{k+1} = \phi_k x_k + \Gamma_k w_k \qquad (3.73)$$

$$z_k = H_k x_k + \varepsilon_k \qquad (3.74)$$

where Equation (3.73) (the state equation) represents the evolution of the parameters in time and Equation (3.74) represents the measurements. The terms w_k and ε_k are discrete time random processes representing state noise, i.e. random inputs in the evolution of the parameters, and measurement errors, respectively. Typically w_k and ε_k are assumed to be independent of each other and uncorrelated from sample to sample. If w_k and ε_k have zero means then it is common to assume that

$$E\{w_k w_j^T\} = Q_k; k = j \qquad (3.75)$$

$$= 0; k \neq j$$

$$E\{\varepsilon_k \varepsilon_j^T\} = R_k; k = j \qquad (3.76)$$

$$= 0; k \neq j$$

The matrices Q_k and R_k are the covariance matrices of the random processes and are allowed to change as k changes.

The matrix φ_k in Equation (3.73) is the state transition matrix. If we imagine sampling a pure sinusoid of the form

$$y(t) = Y_c \cos(\omega t) + Y_s \sin(\omega t)$$

at equal intervals corresponding to $\omega \Delta t = \psi$ then we could take the state to be

$$x_k = \begin{bmatrix} Y_c \\ Y_s \end{bmatrix}$$

and since the state does not change in time

$$\varphi = \begin{bmatrix} 1 & 0 \\ 0 & 1 \end{bmatrix}$$

In this case H_k the measurement matrix would be

$$H_k = [\cos(k\psi)\ \sin(k\psi)]$$

A second possibility is to define the state as

$$x_k = \begin{bmatrix} \cos k\psi & \sin k\psi \\ -\sin k\psi & \cos k\psi \end{bmatrix} \begin{bmatrix} Y_c \\ Y_s \end{bmatrix}$$

so that

$$\varphi_k = \begin{bmatrix} \cos \psi & \sin \psi \\ -\sin \psi & \cos \psi \end{bmatrix}$$

and

$$H_k = [1\ 0]$$

The Kalman filter assumes an initial (before the measurements are made) statistical description of the state x, and recursively (as each measurement becomes available) updates the estimate of state. The initial assumption about the state is that it is a random vector independent of the processes w_k and ε_k and with a known mean and covariance matrix, P_0. The recursive calculation involves computing a gain matrix K_k. The estimate $\hat{x}$ is given by

$$\hat{x}_{k+1} = \varphi_k x_k + K_{k+1}[z_{k+1} - H_{k+1}\varphi_k \hat{x}_k] \tag{3.77}$$

The first term in Equation (3.77) is an update of the old estimate by the state transition matrix while the second is the gain matrix multiplying the observation residual. The bracketed term in Equation (3.77) is the difference between the actual measurement, z_k, and the predicted value of the measurement, i.e. the residual in predicting the measurement. That is, $\phi_k \hat{x}_k$ is the predicted state and $H_{k+1}\phi_k\hat{x}_k$ is the predicted measurement. The computation of the gain matrix K_k involves the computation of two covariance matrices. The first is the covariance of the one-step prediction denoted $P(k+1|k)$. The second is the covariance of the error in the estimate at time k, $P(k|k)$. The first step in computing the gain matrix is to update the covariance for the one-step prediction

$$P(k+1|k) = \phi_k P(k|k)\phi_k^T + \Gamma_k Q_k \Gamma_k^T \tag{3.78}$$

The gain matrix can be computed as

$$K_{k+1} = P(k+1|k)H_{k+1}^T[H_{k+1}P(k+1|k)H_{k+1}^T + R_{k+1}]^{-1} \tag{3.79}$$

The matrix that must be inverted in Equation (3.79) is of the dimension of the measurements which typically is smaller than the number of states. In the line relaying application, the state dimension is 2 while there is only one measurement. In order to maintain the recursion an additional calculation is involved. The matrix $P(k+1|k+1)$ must be formed to be used in the next version of Equation (3.78).

$$P(k+1|k+1) = [I - K_{k+1}H_{k+1}]P(k+1|k) \tag{3.80}$$

Alternate gain expressions are frequently useful. Many forms of Equations (3.78), (3.79) and (3.80) can be obtained using the matrix inversion lemma given in Equation (3.81)

$$(A^{-1} + B^TC^{-1}B)^{-1} = A - AB^T(BAB^T + C)^{-1}BA \tag{3.81}$$

If we write Equation (3.79) in a more compact notation as $K = \mathcal{P}H^T(H\mathcal{P}H^T + R)^{-1}$ and apply the lemma as in Equation (3.82)

$$K = (\mathcal{P}^{-1} + H^TR^{-1}H)^{-1}(\mathcal{P}^{-1} + H^TR^{-1}H)\mathcal{P}H^T(R + H\mathcal{P}H^T)^{-1} \tag{3.82}$$

$$K = (\mathcal{P}^{-1} + H^TR^{-1}H)^{-1}(H^T + H^TR^{-1}H\mathcal{P}H^T)(R + H\mathcal{P}H^T)^{-1}$$

$$K = (\mathcal{P}^{-1} + H^TR^{-1}H)^{-1}H^TR^{-1}(H\mathcal{P}H^T + R)(R + H\mathcal{P}H^T)^{-1}$$

$$K = (\mathcal{P}^{-1} + H^TR^{-1}H)^{-1}H^TR^{-1}$$

Or

$$K_{k+1} = [H^T_{k+1}R_{k+1}H_{k+1} + P^{-1}(k+1|k)]^{-1}H_{k+1}^T R_{k+1} \tag{3.83}$$

Substituting Equation (3.79) into Equation (3.80) with the same notation and employing the lemma again

$$P = \mathcal{P} - \mathcal{P}H^T(H\mathcal{P}H^T + R)^{-1}H\mathcal{P} = (P + H^TR^{-1} + R)^{-1} \tag{3.84}$$

Example 3.22

It seems appropriate to give a comparison of the Kalman estimates and the estimates formed using the DFT algorithms. To do so, however, it is necessary to put the algorithms on a common basis. The Kalman estimate assumes an initial estimate and an initial covariance for that estimate, while the DFT algorithm does not. Equations 3.78–3.80 are not convenient if there is no initial estimate. A form of the equations which is convenient for this case involves the inverse of the covariance matrix P(k|k). If we define the inverse of the covariance matrix as

$$F_k = P(k|k)^{-1}$$

and assume that there is no random input to the system, i.e. $Q_k = 0$, then Equations (3.78) and (3.79) can be rewritten in terms of an update of F_k as

$$F_{k+1} = \varphi_k^{-T} F_k \varphi_k^{-1} + H_{k+1}^T R_{k+1}^{-1} H_{k+1} \tag{3.85}$$

The advantage of Equation (3.82) is that F_0 can be taken as zero. The matrix F_k is referred to as the information matrix and can also be used to compute the gain from

$$F_{k+1} K_{k+1} = H_{k+1}^T R_{k+1}^{-1} \tag{3.86}$$

Equation (3.85) makes it clear that, if the initial F_0 is zero, we must make enough measurements so that F_k is invertible before the gain matrix can be computed. In the relaying problem, this is a manifestation of the fact that one measurement is not sufficient to form an estimate of a phasor. If we take the form of the state equation for the phasor given by $\varphi_k = I$ and $H_k = [\cos(k\psi)\sin(k\psi)]$, and assume that $R_k = I$, the recursion for the information matrix becomes

$$F_{k+1} = F_k + \begin{bmatrix} \cos^2(k+1)\psi & \cos(k+1)\psi\sin(k+1)\psi \\ \cos(k+1)\psi\sin(k+1)\psi & \sin^2(k+1)\psi \end{bmatrix}$$

or

$$F_N = \sum_{k=1}^{N} \begin{bmatrix} \cos^2 k\psi & \cos k\psi \sin k\psi \\ \cos k\psi \sin k\psi & \sin^2 k\psi \end{bmatrix}$$

Under the assumption that $\varphi_k = I$, $Q_k = 0$, $R_k = I$, the recursions are

$$K_k = F_k^{-1} H_k^T \tag{3.87}$$

$$F_k = F_{k-1} + H_k^T H_k \tag{3.88}$$

And using Equation (3.87) in Equation (3.77)

$$\hat{x}_k = \hat{X}_{k-1} + F_k^{-1} H_k^T [z_k - H_k \hat{x}_{k-1}]$$

$$\hat{x}_k = F_k^{-1} [F_k - H_k^T H_k]\hat{x}_{k-1} + H_k^T z_k]$$

and using Equation (3.85)

$$\hat{x}_k = F_k^{-1} [F_{k-1}\hat{x}_{k-1} + H_k^T z_k] \tag{3.89}$$

Equation (3.89) can be repeated to yield

$$\hat{x}_k = F_k^{-1}[F_o x_o + H_1^T z_1 + H_2^T z_2 + \cdots + H_k^T z_k] \qquad (3.90)$$

Equation (3.90) shows that if the initial information matrix is zero then the estimate is formed as the weighted sum of the measurements (if F_0 is zero then the first term in the brackets is zero). The bracketed quantity in Equation (3.87) is

$$\begin{bmatrix} \sum \cos(k\psi)z_k \\ \sum \sin(k\psi)z_k \end{bmatrix}$$

In other words, if there is no initial information and the measurement error has a constant covariance then the Kalman filter estimate is a combination of DFT terms. The particular combination depends on the matrix F_k^{-1}.

If there are an even number of samples per half cycle, at multiples of a half cycle, $\psi = \pi/2n$, $N = 2nm$, i.e. $m = 1$ is a half cycle, $m = 2$ is a full cycle, $m = 3$ is 3 half cycles, and

$$F_N = \sum_{k=1}^{N} \begin{bmatrix} \cos^2(k\psi) & \cos(k\psi)\sin(k\psi) \\ \cos(k\psi)\sin(k\psi) & \sin^2(k\psi) \end{bmatrix} \qquad (3.91)$$

Then using

$$\cos^2 u = \frac{1 + \cos(2u)}{2}, \quad \sin^2 u = \frac{1 - \cos(2u)}{2}, \quad \cos u \sin u = \frac{\sin 2u}{2}$$

With $2u = 2\psi = \pi/n$ and $m = 1$, $N = 2n$, the sums in Equation (3.91) involve a sum of $\cos(\pi/n)$ or $\sin(\pi/n)$ over a full period. Since

$$\sum_{k=1}^{2n} \cos(k\pi/n) = 0, \quad \sum_{k=1}^{2n} \sin(k\pi/n) = 0,$$

$$F_N = \left(\frac{N}{2}\right) I \qquad (3.92)$$

$$\hat{x} = \left(\frac{1}{nm}\right) \begin{bmatrix} \sum_{k=1}^{2nm} \cos(k\pi/2n)z_k \\ \sum_{k=1}^{2nm} \sin(k\pi/2n)z_k \end{bmatrix} \qquad (3.93)$$

In fact, the Kalman filter estimate exactly corresponds to the Fourier estimates in these situations.[11] The estimates given by Equation (3.90) were first derived in a digital relaying context without reference to Kalman filtering. There are seen to be two possible reasons to use the Kalman filter. They are: the existence of an initial estimate, or non-constant measurement noise.

3.10 Summary

In this chapter we have presented the background material needed to understand relaying algorithms. We have examined the Fourier series and Fourier transform as means of determining the frequency content of signals. In a digital relaying context the discrete Fourier transform which takes samples of a signal and computes the various harmonics is of particular importance. The Fourier transform sheds light on the sampling process and explains the need for anti-aliasing filters.

Ideas of probability and random processes are required in order to appreciate how various algorithms estimate the parameters of interest in the relaying application. The view that the solution of an over-defined set of equations is an estimation problem will be used extensively in Chapter 4. The Kalman filter has also been applied in distance relaying. Our presentation of the Kalman filter makes it clear that the primary use of the filter in relaying should be in situations where the measurement noise does not have a constant covariance. When the measurement noise has a constant covariance and when no prior knowledge of the state is assumed, then the Kalman filter solution is equivalent to solutions obtained with other techniques.

Problems

3.1 Determine whether each of the following signals is periodic. If so, give the fundamental frequency.

(a) $r(t) = \cos \omega_o t = \sin \frac{3}{2}\omega_o t$
(b) $r(t) = \cos \pi t + \cos 3t$
(c) $r(t) = \sin \omega_o(t - 3) + \cos \frac{6}{4}\omega_o t$
(d) $r(t) = e^{j\omega_o t} + \sin\omega_o t + \cos \frac{1}{2}\omega_o t$

3.2 Demonstrate the following consequences of Equation (3.9):

(a) If $r(t)$ is real then $c_{-k} = c_k$
(b) If $r(t)$ is even then $c_k = c_{-k}$
(c) If $r(t)$ is odd then $c_k = -c_{-k}$
(d) If $r(t)$ is real and even then $c_k = c_{-k} = c_k^*$ i.e. all c_k are pure real
(e) If $r(t)$ is real and odd then $c_k = -c_{-k} = -c_k^*$ i.e. all c_k are pure imaginary

3.3 Let $T > 0$ be a fixed number

(a) Let $x(t)$ be a signal which is zero for $t \leq 0$ and $t \geq T$. Show that the following infinite sum defines a periodic signal, $r(t)$

$$r(t) = \sum_{n=-\infty}^{n=\infty} x(t - nT)$$

(b) Let $x(t)$ be a signal which is zero for $t \leq 0$ and $t \geq T/2$. Construct an even signal $r(t)$ such that
    i. $r(t)$ is periodic
    ii. $r(t) = x(t)$ for $0 \leq t \leq T/2$

(c) Repeat (b) for an odd signal

3.4 (a) Find the Fourier series coefficients $\{c_k\}$ for the signal

$$r(t) = |\cos\Omega t|; \ \Omega > 0 \ -\infty < t < \infty$$

(b) Write the Fourier series for $r(t)$ in complex exponential form and in sine and cosine form.

3.5 Given that an LTI system has an impulse response

$$h(t) = u(t)e^{-t}\cos t$$

Find the response of the system to the input

$$w(t) = \sin t; \ -\infty < t < \infty$$

3.6 An LTI system is known to have impulse response

$$h(t) = u(t)e^{-4t}$$

(a) Determine the response, $y(t)$, of the system to the input $w(t) = e^{j\omega t}$. Sketch the magnitude of the response as a function of ω.

(b) Find the Fourier series coefficients of the output if the input is $r(t)$ from 3.3

(c) Is the output an even function of time?

3.7 Let $r(t)$ be a periodic signal with a fundamental period T_o and fundamental frequency of $\omega_o = 2\pi/T_o$ Let c_k denote the kth Fourier series coefficient of $r(t)$. Find formulas in terms of c_k for the kth Fourier coefficients d_k or $y(t)$ where

(a) $y(t) = r(-t)$
(b) $y(t) = r(t - t_1)$

(c) $y(t) = \dfrac{dr}{dt}$ (assuming $r(t)$ is differentiable)

3.8 Find the Fourier transforms $\hat{X}(\omega)$ of each signal x(t)

(a) $x(t) = \begin{cases} \cos t & |t| \leq 1 \\ 0 & |t| \rangle 1 \end{cases} = P_1(t) \cos t$

(b) $x(t) = e^{-|t|} P_1(t)$

(c) $x(t) = \dfrac{1}{a} P_{\frac{a}{2}}(t)$

3.9 An LTI system designed to 'smooth' input signals has the following output when w(t) is the input

$$y(t) = \int_{t-a/2}^{t+a/2} w(\tau)\, d\tau \quad -\infty \langle t \langle \infty$$

Find the frequency response of the system.

3.10 Consider the signal

$$x(t) = \sin t; \quad |t| < \pi$$

$$0; \quad |t| > \pi$$

(a) Find $\hat{X}(\omega)$ by direct integration
(b) let $x(t) = \sin t \, p_\pi(t)$ where

$$p_\pi(t) \overset{F}{\leftrightarrow} \frac{2 \sin \pi \omega}{\omega}$$

Find $\hat{X}(\omega)$ using the modulation rule.

3.11 Let $x(t) = e^{-\frac{1}{2}\left(\frac{t^2}{T}\right)}$

(a) Show that $\frac{d}{d\omega}[\hat{X}(\omega)] = -\omega T^2 \hat{X}(\omega)$ and conclude that $\hat{X}(\omega) = Ke^{-\left(\frac{\omega^2 T^2}{2}\right)}$ for some constant K.

(b) Given that $\int_{-\infty}^{\infty} e^{-\tau^2} d\tau = \sqrt{\pi}$, show that $K = T\sqrt{2\pi}$

3.12 Recall that given a signal, x(t), we may write x(t) uniquely as the sum of an even signal $x_e(t)$ and an odd signal $x_o(t)$

$$x_e(t) = \frac{1}{2}[x(t) + x(-t)]$$

$$x_o(t) = \frac{1}{2}[x(t) - x(-t)]$$

(a) If x(t) is real-valued and has Fourier transform $\hat{X}(\omega)$, show that

$$\text{Re}\{\hat{X}(\omega)\} = \hat{X}_e(\omega)$$

$$\text{Im}\{\hat{X}(\omega)\} = -j\hat{X}_o(\omega)$$

(b) Suppose that x(t) is causal, i.e. x(t) = 0 for t< 0. Assume that F^{-1} exists for $\hat{X}_e(\omega)$ and $\hat{X}_o(\omega)$. Show that

$$x(t) = \frac{2}{\pi} \int_0^\infty \text{Re}\{\hat{X}(\omega)\} \cos \omega t \, d\omega; \, t \rangle \, 0$$

$$x(t) = \frac{2}{\pi} \int_0^\infty \text{Im}\{\hat{X}(\omega)\} \sin \omega t; \, t \rangle \, 0$$

that is, a causal signal may be determined from either the real or imaginary part of its Fourier transform.

3.13 Suppose that a certain linear time-invariant system has response

$$y(t) = [6e^{-4t} - 6e^{-5t}] u(t)$$

to the input

$$w(t) = [3e^{-5t} - 2e^{-4t}] u(t)$$

Find the frequency response of the system.

3.14 (a) Find the impulse response of the linear time-invariant system whose frequency response is

$$\hat{H}(\omega) = A_o e^{-\alpha\omega^2} e^{-j\omega t_1}; \, -\infty \, \langle \, \omega \, \langle \, \infty$$

this system might be called an ideal linear phase Gaussian filter.
(b) Is the system causal?

3.15 The signal shown in Figure P3.15a has Fourier transform given by

$$\hat{X}_o(\omega) = \frac{4 \sin^2 \omega}{\omega^2}$$

without integrating, find the transforms of the following signals:

(a) The signal shown in Figure P3.15b
(b) $x_2(t) = t x_o(t)$
(c) $x_3(t) = x_o(t) \sin \omega_o t$

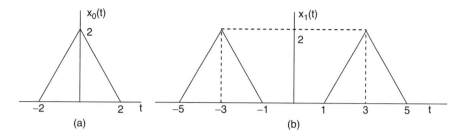

Figure P3.34   Signals for Problem 3.15

3.16 Find the transforms of

(a) $x(t) = \dfrac{4 \sin^2 t}{t^2}$

(b) $x(t) = \dfrac{1}{jt} + \pi \delta(t)$

3.17 Suppose $x_1(t)$ and $x_2(t)$ are signals such that

$$\hat{X}_1(\omega) = 0 \; |\omega| \rangle \; \omega_1$$

$$\hat{X}_2(\omega) = 0 \; |\omega| \rangle \; \omega_2(\omega_2 \rangle \; \omega_1)$$

Find the minimum sampling rate necessary to represent $y(t)$ exactly, when

(a) $y(t) = x_1(t) + x_2(t)$
(b) $y(t) = x_1(t) \, * \, x_2(t)$
(c) $y(t) = x_1(t)x_2(t)$

3.18 Recovering an approximation of a signal $x(t)$ from a sampled version may be viewed as a process of interpolation. That is, given $x(nT)$, $-\infty < n < \infty$, T fixed, find an approximation $\tilde{x}(t)$ to $x(t)$, where

$$\tilde{x}(t) = \sum_{n=-\infty}^{n=\infty} x(nT)h(t - nT)$$

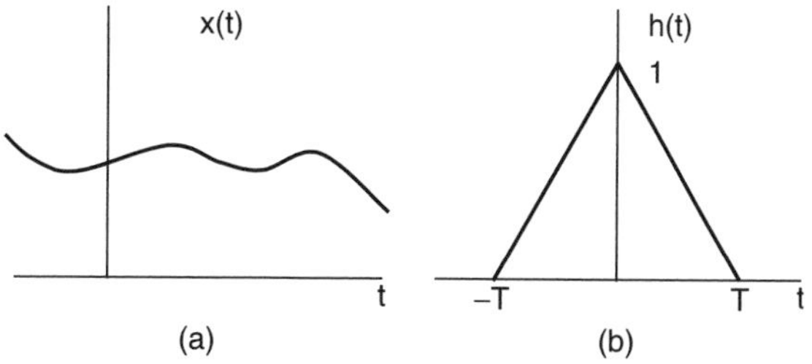

Figure P3.35   (a) A signal to be interpolated. (b) h(t) for linear interpolation

where h(t) is a function which goes to zero as $|t| \to \infty$. Let x(t) be as shown in Figure P3.18a.

(a) Sketch the zero-order hold interpolation $\tilde{x}(t)$ when $h(t) = p_{T/2}(t)$.
(b) Sketch the linear interpolation $\tilde{x}(t)$ when h(t) is as shown in P3.18b.

References

[1] Papoulis, A. *The Fourier Integral and its Application*, McGraw-Hill, 1961.
[2] Oppenheim, A.V. and Willsky A.S. (1983) *Signals and Systems*, Prentice-Hall.
[3] Lathi, B.P. (1987) *Signals and Systems*, Berkeley-Cambridge.
[4] Papoulis, A. (1977) *Signal Analysis*, McGraw-Hill.
[5] Rabiner, L.R. and Gold, B. (1975) *Theory and Application of Digital Signal Processing*, McGraw-Hill.
[6] Openheimer, A.V. and Schaefer, R.W. (1975) *Digital Signal Processing*, Prentice-Hall.
[7] Helstrom, C.W. (1984) *Probability and Stochastic Processes for Engineers*, Macmillian.
[8] Anderson, B.D.O. and Moore, J.B. (1979) *Optimal Filtering*, Prentice-Hall.
[9] Gelb, A. (1974) *Applied Optimal Estimation*, MIT Press.
[10] Meditch, J.S. (1969) *Stochastic Optimal Linear Estimation and Control*, McGraw-Hill.
[11] Thorp, J.S., Phadke, A.G., Horowitz, S.H. and Beehier, J.E. (1979) Limits to impedance relaying, IEEE Trans. on PAS, vol. 98, no. 1, pp. 246–260.

4

Digital filters

4.1 Introduction

The material in Chapter 3 concerning Fourier transforms, random processes, and analog filters is primarily cast in continuous time. Much of the subsequent literature has been focused on the application of techniques from digital signal processing. This chapter parallels Chapter 3 with an emphasis on some of these topics, including: discrete time signals and systems, the design of digital filters and windowing. The chapter concludes with wavelet transforms, artificial neural networks, and an introduction to decision trees and agents. Some of the later ideas have yet to be applied on a large scale but they represent a dramatic shift from the early reluctance to use computers in protection to an embrace of many new approaches to these problems.

4.2 Discrete time systems

The processes and operations that digital relays perform on the samples of voltage and current can be regarded as manifestations of discrete time systems. To understand such systems we need first to examine the notation and definitions that are used with discrete time signals. As in Section 3.6, we will use the notation x[n] for a function with integer arguments.

Example 4.1

The discrete time step and impulse are given by

$$x[n] = u[n] = \begin{Bmatrix} 1; & n \geq 0 \\ 0; & n < 0 \end{Bmatrix} \tag{4.1}$$

$$x[n] = \delta[n] = \begin{Bmatrix} 1; & n = 0 \\ 0; & n \neq 0 \end{Bmatrix} \tag{4.2}$$

Note that $\delta[n] = u[n] - u[n-1]$
A discrete time exponential is

$$x[n] = z_0^n$$

Where z_0 is any complex number. Then $|z_0^n|$ grows for $|z_0| > 1$ and decays for $|z_0| < 1$ and if

$$x[n] = e^{j\Omega} \tag{4.3}$$

then $x[n]$ is a discrete time sinusoid. Discrete time signals are causal if they are zero for negative arguments, i.e. $x[n] = 0$, $n < 0$ and time limited if they are confined to an interval, i.e. for some $n_1 < n_2$

$$x[n] = 0 \quad n < n_1 \quad \text{and} \quad n > n_2 \tag{4.4}$$

Just as in Section 3.2, a discrete time signal is periodic if

$$x[n] = x[n+N]$$

for some N.

4.2.1 Operations on discrete time sequences

Given any two discrete time sequences $x_1[n]$ and $x_2[n]$ the following operations are defined

$$z[n] = x_1[n] + x_2[n] \text{ addition} \tag{4.5}$$

$$z[n] = cx_1[n] \text{ multiplication by a scalar}$$

$$z[n] = x_1[n] x_2[n] \text{ multiplication of sequences}$$

$$z[n] = x[n - n_0] \text{ time shift}$$

4.2.2 Convolution

An important operation involving two sequences is discrete time convolution $x_1[n]*x_2[n]$ defined by

$$x_1[n]*x_2[n] = \sum_{k=-\infty}^{\infty} x_1[k]x_2[n-k] \tag{4.6}$$

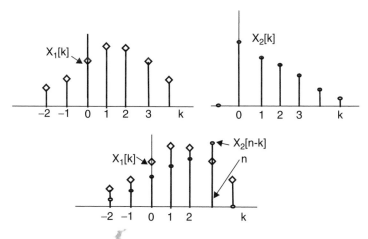

Figure 4.1   Discrete time convolution

Equation (4.6) is the discrete equivalent of continuous time operation is Section 3.5. The terms inside the summation are shown in Figure 4.1. The sequence $x_1[k]$ is fixed while the sequence $x_2[n-k]$ is $x_2[k]$ time reversed and shifted so the origin is at $k = n$.

Example 4.2   Convolution of a step with a step

$$x_1[n] = x_2[n] = u[n]$$

$$u[n]^*u[n] = \sum_{k=-\infty}^{\infty} u[k]u[n-k]$$

$$u[n]^*u[n] = \sum_{k=0}^{\infty} u[n-k]$$

$$u[n]^*u[n] = \sum_{k=0}^{n} 1 \, n \geq 0$$

$$= 0 \qquad n < 0$$

$$u[n]^*u[n] = (n+1)\, u[n]$$

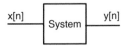

Figure 4.2   Discrete time system

4.3   Discrete time systems

A discrete time system operates on an input discrete time sequence to produce another discrete time sequence as an output, as shown in Figure 4.2. The system is linear if for any pair of inputs $y_1[n]$ and $y_2[n]$ with corresponding outputs $x_1[n]$ and $x_2[n]$, the input $y_1[n] + y_2[n]$ produces the output $x_1[n] + x_2[n]$. The system is time invariant if for any input $y[n]$ with output $x[n]$, the input $y[n - n_0]$ produces the output $x[n - n_0]$. The system is causal if the output $x[n]$ depends only on the input $y[m]$ for $m \leq n$. Just as in continuous time, convolution with an impulse reproduces the signal, i.e.

$$y[n]*\delta[n] = \sum_{k=-\infty}^{\infty} y[k]\delta[n - k] = y[n] \tag{4.7}$$

Combining the facts in the preceding paragraph we can characterize a linear time invariant system by its response to a unit impulse. If the response of a linear time invariant system to a unit impulse $\delta[n]$ is $h[n]$ as shown in Figure 4.3 then the response to any input $y[n]$ is $x[n] = y[n]*h[n]$. This can be established by writing the input $y[n]$ as a convolution with an impulse

$$y[n] = \sum_{k=-\infty}^{\infty} y[k]\delta[n - k] \tag{4.8}$$

The output $x[n]$ then, using linearity, is the sum of the responses to the shifted impulses inside the sum or

$$x[n] = \sum_{k=-\infty}^{\infty} y[k]h[n - k] \tag{4.9}$$

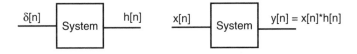

Figure 4.3   Discrete time impulse response

Note that the time invariance property is used in writing $h[n - k]$ as the response to $\delta[n - k]$.

Discrete time systems fall into two large classes, the first is those in which the impulse response is limited in duration so that the output only depends on a finite number of previous inputs as in Equation (4.10).

$$x[n] = a_0 y[n] + a_1 y[n - 1] + \cdots + a_k y[n - k] \tag{4.10}$$

The second is those for which the output depends on all previous input values. The first are designated Finite Impulse Response (FIR) while the second are designated as Infinite Impulse Response (IIR). If we cast a relaying algorithm as a discrete time system or a digital filter it will be an FIR filter since the number of past samples of the input that can be used in making a decision must be finite. We will encounter IIR filters, however, in examining equivalences between continuous time and discrete time filters.

4.4 Z Transforms

4.4.1 Power series

Consider a power series of the form of Equation (4.11) where w is a complex variable.

$$f(w) = \sum_{n=-\infty}^{n=\infty} c_n w^n \tag{4.11}$$

The series will converge for some values of w. A simple example is given in Equation (4.12).

$$f(w) = \sum_{n=0}^{n=N} w^n = 1 + w + w^2 + \cdots w^N \tag{4.12}$$

$$(1 + w + w^2 + \cdots w^N)(1 - w) = 1 + w + w^2 + \cdots w^N - (w + w^2 + \cdots w^{N+1})$$
$$= 1 - w^{N+1}$$

or

$$f(w) = \sum_{n=0}^{n=N} w^n = \frac{1 - w^{N+1}}{1 - w} \tag{4.13}$$

Being a finite sum, the series in Equation (4.12) converges for all w. The expression in Equation (4.13) must be evaluated carefully at $w = 1$ but use of L'Hospital's rule gives $f(1) = N + 1$ which is obviously correct. If we let $N \to \infty$ in Equation (4.13) it is necessary that $|w| < 1$ for the series to converge In general the sum in form of Equation (4.14) converges in a disk given by Equation (4.15) which is referred to as the ratio test.

$$f(w) = \sum_{n=0}^{\infty} c_n w^n \qquad (4.14)$$

$$|w| < R < \lim_{n \to \infty} \left| \frac{c_n}{c_{n+1}} \right| \qquad (4.15)$$

If $R = 0$ the sum does not converge for any value of w. R is called the radius of convergence.

Example 4.3 Ratio tests

(a) $f_a(w) = \sum_{n=0}^{\infty} n w^n; \qquad R = \lim_{n \to \infty} \frac{n}{n+1} = 1$

(b) $f_b(w) = \sum_{n=0}^{\infty} \frac{1}{n!} w^n; \qquad R = \lim_{n \to \infty} \frac{(n+1)!}{n!} = \infty$

In fact, $f_b(w) = e^z$ converges everywhere.

4.4.2 Z Transforms

The z transform is the discrete time equivalent of the Laplace transform and is given by Equation (4.16) for the sequence $h[n]$

$$H(z) = \sum_{n=-\infty}^{\infty} h[n] z^{-n} \qquad (4.16)$$

If the transform exists it is defined in an annual region referred to as the region of convergence in the z-plane defined by

$$R_- < |z| < R_+ \qquad (4.17)$$

Note the negative exponent on z in Equation (4.16). Letting $z = 1/w$ in the ratio test in preceding section establishes that R+ comes from the limit as $n \to \infty$ and R- comes from the limit as $n \to -\infty$.

Example 4.4 Z Transforms

(a) The unit step $h[n] = 1$, $n = 0, 1, 2\ldots$(using Equation 4.13 and $z = 1/w$)

$$H(z) = \sum_0^\infty z^{-n} = \frac{1}{1 - (1/z)} = \frac{z}{z - 1}; \ |z| < 1$$

(b) The discrete unit ramp $h[n] = nu[n]$ (differentiating 4.4a with respect to z and multiplying by $-z$)

$$-\sum_0^\infty nz^{-n-1} = \frac{1}{z - 1} - \frac{z}{(z - 1)^2} \qquad \sum_0^\infty nz^{-n} = \frac{-z}{z - 1} + \frac{z^2}{(z - 1)^2}$$

$$H(z) = \sum_0^\infty nz^{-n} = \frac{z}{(z - 1)^2}; \ |z| < 1$$

(c) A sampled exponential $h[n] = u[n]e^{-anT}$ (using Equation (4.13) and $ze^{aT} = (1/w)$)

$$H(z) = \sum_0^\infty e^{-aTn}z^{-n} = \sum_0^\infty (e^{-aT}z^{-1})^n = \frac{ze^{aT}}{ze^{aT} - 1} = \frac{z}{z - e^{-aT}}; \ |z| < |e^{-aT}|$$

If $\beta = e^{-aT}$, $h[n] = u[n]\beta^n$ and $H(z) = \frac{z}{z-\beta}$; $|z| < |\beta|$

4.4.3 Inverse Z transforms

The inverse z transform is given by Equation 4.18 where the integration is along a counterclockwise curve in the region of convergence.

$$h[n] = \frac{1}{2\pi j} \oint_C H(z)z^{n-1} \, dz \qquad (4.18)$$

Equations (4.16) and (4.18) constitute a z transform pair. The ROC in Equation (4.17) and the path of integration in Equation (4.18) are an important part of the relationship.

For rational functions, especially with simple pole, the use of partial fractions offers a convenient alternative to the integration in Equation (4.18). Given a rational function, H(z), with distinct poles and numerator of lower degree than the denominator write a partial fraction expansion for H(z)/z

$$\frac{H(z)}{z} = \sum_k \frac{\alpha_k}{(z - \beta_k)}; \quad \text{where} \quad \alpha_k = (z - \beta_k)\frac{H(z)}{z}|_{z \to \beta_k} \tag{4.19}$$

Repeated poles and numerators of higher order than the denominator can be accommodated as in Laplace transforms. The division by z before the partial fraction expansion gives the desired form for the expansion of H(z).

$$H(z) = \sum_k \frac{\alpha_k z}{(z - \beta_k)}; \quad h[n] = \sum_k \alpha_k \beta_k^n \tag{4.20}$$

4.4.4 Properties of Z transforms

With the notation as in Chapter 3 for Fourier transforms shown in Equation (4.21) the properties of z transforms have direct counterparts with the properties of the Fourier transform in Chapter 3.

$$h[n] \overset{z}{\leftrightarrow} H(z) \tag{4.21}$$

P1) Linearity
If

$$h_1[n] \overset{z}{\leftrightarrow} H_1(z)$$

and

$$h_2[n] \overset{z}{\leftrightarrow} H_2(z)$$

then

$$c_1 h_1[n] + c_2 h_2[n] \overset{z}{\leftrightarrow} c_1 H_1(z) + c_2 H_2(z) \tag{4.22}$$

P2) Time shifting
If

$$h_1[n] \overset{z}{\leftrightarrow} H_1(z)$$

then

$$h_1[n - n_0] \overset{z}{\leftrightarrow} z^{-n_0} H_1(z) \tag{4.23}$$

P3) Differentiation in z
If

$$h_1[n] \overset{z}{\leftrightarrow} H_1(z)$$

then

$$(n-1)h_1[n-1] \overset{z}{\leftrightarrow} -\frac{d}{dz}H_1(z)$$

or from time shifting

$$nh_1[n] \overset{z}{\leftrightarrow} -z\frac{d}{dz}H_1(z) \tag{4.24}$$

P4) Convolution
   If

$$h_1[n] \overset{z}{\leftrightarrow} H_1(z)$$

and

$$h_2[n] \overset{z}{\leftrightarrow} H_2(z)$$

then

$$h_1[n]^*h_2[n] \overset{z}{\leftrightarrow} H_1(z)H_2(z) \tag{4.25}$$

Note that, since we have established that a linear discrete time system has an output which is the convolution of its impulse response with the input sequence property, P4 says the x transform of the output is the product of the z transform of the input and the z transform of the impulse response.

$$x[n] = h[n]^*y[n] \overset{z}{\leftrightarrow} H(z)Y(z) = X(z) \tag{4.26}$$

The z transform of the impulse response, $H(z)$, is then the discrete time transfer function.

Example 4.5 Solution of a constant coefficient difference equation

Property P2 enables the z transform to be used in the solution of constant coefficient difference equations in the same way that Laplace transforms are used to solve differential equations. The solution of the differential equation is somewhat more of an accomplishment, as can be seen considering the difference equation in Equation (4.27) relating the input sequence $y[n]$ and output sequence $x[n]$.

$$x[n] + (5/4)x[n-1] + (3/8)x[n-2] = y[n-1] + 4y[n-2] \tag{4.27}$$

Taking z transforms and using P2 and partial fractions

$$X(z) + (5/4)z^{-1}X(z) + (3/8)z^{-2}X(z) = z^{-1}Y(z) + 4z^{-2}Y(z)$$

$$X(z)[1 + (5/4)z^{-1} + (3/8)z^{-2}] = (z^{-1} + 4z^{-2})Y(z)$$

$$X(z) = \frac{(z+4)Y(z)}{(z^2 + (5/4)z + (3/8))} = \frac{(z+4)Y(z)}{(z+1/2)(z+3/4)}$$

If y[n] is a unit step and the initial condition are $x[n] = 0$ for $n \leq 0$

$$X(z)/z = \frac{(z+4)}{(z+1/2)(z+3/4)(z+1)}$$

$$X(z)/z = \frac{-28/3}{z+1/2} + \frac{52/7}{z+3/4} + \frac{(40/21)}{z-1}$$

$$x[n] = (-28/3)(-1/2)^n + (52/7)(-3/4)^n + (40/21)$$

For comparison the difference Equation can be simply rewritten as

$$x[n] = (-5/4) \times [n-1] - (3/8)x[n-2] + u[n-1] + 4u[n-2]$$

4.4.5 Discrete time fourier transform

In the special case that the unit circle is in the ROC, the z transform can be used to create the discrete time Fourier transform by letting $z = e^{j\Omega}$. The relationships are in Equation (4.28).

$$\hat{X}(\Omega) = \sum_{-\infty}^{\infty} x[n]e^{-jn\Omega}$$

$$x[n] = \frac{1}{2\pi j} \int_{-\pi}^{\pi} \hat{X}(\Omega)e^{jn\Omega}\, d\Omega \tag{4.28}$$

If $X(z)$ is the z transform then $X(e^{j\Omega}) = \hat{X}(\Omega)$. Note that $\hat{X}(\Omega)$ is periodic with period 2π and the integral in Ω can be performed over any interval of 2π. The interval $[-\pi, \pi]$ is chosen only to make the parallel to the Fourier transform in continuous time. The discrete samples in time have produced a transform which is continuous in Ω but periodic. Consider the ideal low-pass filter shown in Figure 4.4.

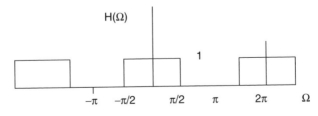

Figure 4.4 Discrete time ideal low-pass filter

The inverse discrete time Fourier transform is

$$x[n] = \frac{1}{2\pi} \int\limits_{-\pi/2}^{\pi/2} \hat{X}(\Omega) e^{jn\Omega} \, d\Omega = \frac{e^{jn\pi/2} - e^{-jn\pi/2}}{j2n\pi} = \frac{\sin(n\pi/2)}{n\pi}$$

Figures 3.9 and 3.10 are the continuous time equivalents with $\omega t = n\pi/2$.

Since the z transform is a subset of the z transform all of the properties of the z transform including convolution carry over. Hence we can imagine specifying the behavior of a discrete time system in terms of its frequency response because a filter with impulse response h[n] will operate on an input y[n] to produce an output x[n] with

$$\hat{X}(\Omega) = \hat{H}(\Omega)\hat{Y}(\Omega) \tag{4.29}$$

Excepting the fact that $\hat{H}(\Omega)$ is periodic all the nomenclature associated with analog filters is appropriate for digital filters, $\hat{H}(\Omega)$ can be low-pass, high-pass or band-pass. Digital filters have pass bands, transition bands, and stop bands.

4.5   Digital filters

One approach to digital filter design is to make the impulse response of the digital filter correspond to samples of the analog filter's impulse response. Such a filter is referred to as 'impulse invariant'. Sampling as in Section 3.5 with samples spaced at T_0 seconds and using an infinite impulse train as the sampling function the transform of the sampled function is given by Equation (4.30).

$$z(t) = x(t) \sum_{-\infty}^{\infty} \delta(t - nT_o) \overset{\mathscr{F}}{\leftrightarrow} Z(\omega) = \frac{1}{T} \sum_{-\infty}^{\infty} X(\omega - k\omega_0) \tag{4.30}$$

where $\omega_0 = 2\pi/T_0$. The function z(t) is a continuous time function made up of impulses with weights given by the sample of x at nT_0. Taking the transform of the impulses $Z(\omega)$ is also given by Equation (4.31).

$$Z(\omega) = \sum_{-\infty}^{\infty} x(nT_o) e^{-jn\omega T_0} \tag{4.31}$$

Comparing Equations (4.28) and (4.31)

$$\hat{X}(\Omega) = Z(\Omega/T_0) \tag{4.32}$$

The periodic function $\hat{X}(\Omega)$ is simply periodic replicas of the original frequency response. If $X(\omega)$ is band limited and T_0 is chosen to satisfy the Nyquist criteria then the discrete filter is impulse invariant with the continuous time filter.

If the analog filter were a second order Butterworth filter, for example, several difficulties emerge. The frequency response in not band limited and there is aliasing at any reasonable sampling rate and the digital filter is IIR. Compromises can be made to try to minimize the effects of both of these problems but other approaches seem needed.

The mapping $z = e^{j\Omega}$ limits the complex variable z to the unit circle. It can be viewed as the mapping $z = e^{sT}$ with $s = j\Omega/T_0$. Other such mappings include the bilinear transformation

$$s = \frac{2}{T_0} \frac{1 - z^{-1}}{1 + z^{-1}} \tag{4.33}$$

which has an inverse

$$z = \frac{1 + (T_0/2)s}{1 - (T_0/2)s} \tag{4.34}$$

For $z = e^{j\Omega}$ the pair of equations are given in Equation (4.35)

$$\omega = \frac{2}{T_0} \tan \frac{\Omega}{2} \quad \text{and} \quad \Omega = 2 \tan^{-1} \frac{\omega T_0}{2} \tag{4.35}$$

Given an analog filter with transfer function $H_a(s)$, the discrete time transfer function is given by Equation (4.36)

$$H_d(z) = H_a \left(\frac{2}{T} \frac{1 - z^{-1}}{1 + z^{-1}} \right) \tag{4.36}$$

Aliasing is eliminated but the mapping distorts the frequency axis which can have an effect on some designs. The distortion is shown in Figure 4.5

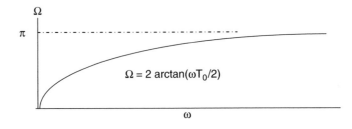

Figure 4.5 Frequency distortions in the bilinear transformation

4.6 Windows and windowing

Windowing is an issue in spectral estimation, that is, estimating the frequency content of a signal record and in filter design. The FFT introduced in Section 3.6 is a convenient tool for spectral analysis. Recall: the FFT is an efficient computation technique for computing the terms in the DFT. The DFT assumes that the signal record in question is one period of a periodic function and provides a connection between N samples in time and N complex numbers which describe the harmonic content of the signal. In practical applications, the record is unlikely to be one period of a periodic signal and the assumption is that the harmonic content found is that of the periodic extension of the signal record as shown in Figure 4.6. The fact that the first and last samples are not identical produces a discontinuity in the periodic extension and alters the frequency content. The sample values of the waveform in Figure 4.6 are

$$x = [-0.3647 \ -0.2147 \ -0.0647 \ 0.0853 \ 0.2353 \ 0.3853 \ 0.5353 \ 0.6353 \ 0.6203$$

$$0.5853 \ 0.4353 \ 0.2353 \ -0.0647 \ -0.5647 \ -1.1647 \ -1.3147]$$

The absolute value of the DFT with a rectangular window and with a Hamming window are shown in Figure 4.7. It can be seen that the window has greatly reduced the energy in the all the harmonics while increasing the energy at DC. Viewed as a filter itself, the Hamming window is shown in Figure 4.8 compared to the rectangular window. The Hamming response is broader around the fundamental (it does not reject the second harmonic) but is much more selective for higher frequencies. There are a large number of windows to choose from. Matlab offers 16 different windows. National Instruments offers a number in LabView.[1] Most popular windows have the features of the Hamming window to varying degrees. Some have broader main lobes and greater rejection of side lobes. Some have adjustable parameters that control

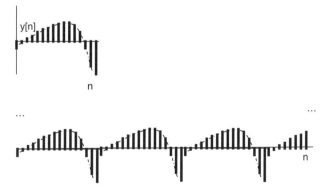

Figure 4.6 A signal and its periodic extension

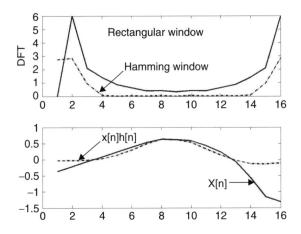

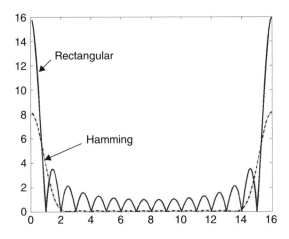

Figure 4.7   Effect of a Hamming window

Figure 4.8   The Hamming and rectangular window transfer functions

these effects. The choice of a windowing functions is complicated and depends on the application.

The expressions for the window function for three representative windows are given in Table 4.1. The 32 sample windows and the corresponding FFTs are shown in Figure 4.9.

4.7 Linear phase

One of the features of the windowing functions in the previous section is that they come in at least one form with linear phase. Linear phase in analog filters is an unachievable ideal while it is routine in digital filters. The elimination of phase distortion is important in many applications. An FIR filter with certain symmetries

Table 4.1 Three windows

Hamming[2]

$$w[n] = 0.54 - 0.46 \cos\left(\tfrac{2\pi}{N}n\right), \ 0 \le n \le N$$

2

4-term Blackman-Harris[3]

$$w[n] = 0.35875 + 0.48829 \cos\left(\tfrac{2\pi}{N}n\right) + 0.01168 \cos\left(\tfrac{2\pi}{N}3n\right)$$

$$\text{for } -\tfrac{N}{2} \le n \le \tfrac{N}{2}$$

Flat-top[4]

$$w[n] = \alpha_0 + \alpha_1 \cos\left(\tfrac{2\pi}{N}n\right) + \alpha_2 \cos\left(\tfrac{2\pi}{N}2n\right) + \alpha_3 \cos\left(\tfrac{2\pi}{N}3n\right) + \alpha_4 \cos\left(\tfrac{2\pi}{N}4n\right)$$

where $\alpha = [0.215578 - 0.416633 \ 0.277263 - 0.083579 \ 0.006947]$

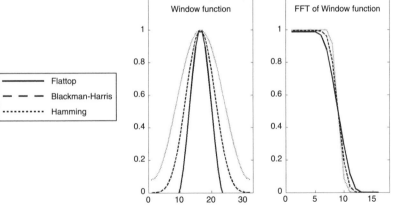

Figure 4.9 Window functions and their FFTs

has linear phase. Let h[n] be the impulse response of an FIR filter with

$$\hat{H}(\Omega) = \overline{H}(\Omega)e^{i\theta(\Omega)} \tag{4.37}$$

with $\overline{H}(\Omega)$ real and

$$h[n] = 0 \quad \text{for} \quad n < 0, \quad n \ge N \tag{4.38}$$

If

$$h[n] = h[N - 1 - n] \tag{4.39}$$

then

$$\theta(\Omega) = -\left(\frac{N-1}{2}\right)\Omega \tag{4.40}$$

Assume N is even, (a similar development is possible if N is odd) and group the terms in the sum

$$\hat{H}(\Omega) = \sum_0^{N/2-1} h[n](e^{-jn\Omega} + e^{-j(N-1-n)\Omega})$$

$$\hat{H}(\Omega) = \sum_0^{N/2-1} h[n]e^{-j\left(\frac{N-1}{2}\right)\Omega} \left(e^{-j\left(n-\left(\frac{N-1}{2}\right)\right)\Omega} + e^{-j\left(N-1-n-\left(\frac{N-1}{2}\right)\right)\Omega}\right)$$

$$\hat{H}(\Omega) = \sum_0^{N/2-1} h[n]e^{-j\left(\frac{N-1}{2}\right)\Omega} \left(e^{-j\left(n-\left(\frac{N-1}{2}\right)\right)\Omega} + e^{j\left(n-\left(\frac{N-1}{2}\right)\right)\Omega}\right)$$

$$\hat{H}(\Omega) = e^{-j\left(\frac{N-1}{2}\right)\Omega} \sum_0^{N/2-1} 2\, h[n] \cos\left(n - \left(\frac{N-1}{2}\right)\Omega\right) \tag{4.41}$$

For N odd, the upper limit on the sum is $(N/2 - 1)$. A similar result is obtained with odd symmetry (a minus sign in Equation 4.38), i.e.,

$$h[n] = -h[N - 1 - n] \tag{4.42}$$

With odd symmetry the cosine terms in the sum in Equation (4.41) are replaced with sine terms but the angles $\theta(\Omega)$ is the same.

4.8   Approximation – filter synthesis

The design of a digital filter can be initiated by specifying requirements on the frequency response $\hat{H}(\Omega)$. Generic specifications for a low-pass filter are shown in Figure 4.10.

$$1 - \delta_1 \le \hat{H}(\Omega) \le 1 + \delta_1 \quad 0 \le \Omega \le \Omega_p$$
$$0 \le \hat{H}(\Omega) \le \delta_2 \quad \Omega_s \le \Omega \le \pi$$

It is expected that the frequency response will have ripples in the pass band and stop band. One approach would be to window a portion of the impulse response of the ideal low-pass filter. Let g[n] be given by Equation (4.43)

$$g[n] = \frac{\sin \Omega_c(n - M)}{\pi(n - M)} \quad 0 \le n \le 2M \tag{4.43}$$

Then by windowing g[n] with various windows, different low-pass filters are obtained. Figure 4.11 shows the frequency response of a 32 (M = 16) sample version of Equation (4.43) with a cut-off frequency of $\pi/2$ along with the effect of

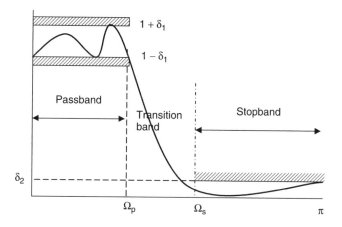

Figure 4.10   Magnitude response specifications for a low-pass filter

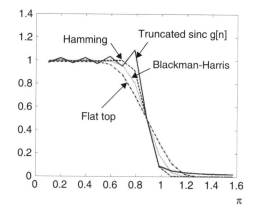

Figure 4.11   Low-pass filters $N = 32$

the Hamming, Flat top, and Blackman-Harris windows. All four filters have linear phase since they are symmetric.

A more direct approach to digital filter design would be to approach the choice of the impulse response coefficients as an optimization problem. That is, from the filter specification in Figure 4.10, for a given stop band and pass band find a set of filter coefficients of given order to minimize the maximum deviation from 1 in the pass band and 0 in the stop band. It is even possible to assign different weights to the two different types of error. The difficulty is that algorithms for minimizing maximum deviations are more complicated than for many other minimization problems. The Parks McClellan algorithm uses the Remez exchange algorithm to solve these problems.[5] The function firpm.m in MATLAB is an implementation of that algorithm. The FFT of the 32d order filter with the indicated pass and stop bands is shown in Figure 4.12.

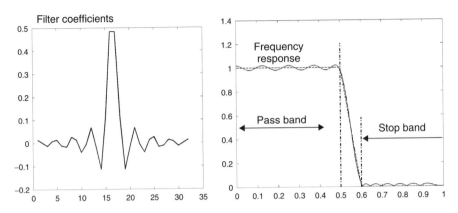

Figure 4.12 Parks McClellan optimum 32nd order low-pass filter with e = 0.0197

4.9 Wavelets

The wavelet transform is a signal processing tool that has replaced the Fourier transform in many applications including data compression, Sonar and Radar, communications and biomedical applications. Considerable overlap exists between wavelets and the area of filter banks. The wavelet transform is viewed as an improvement over the Fourier transform because it deals with time-frequency resolution in a different way. The Fourier transform provides a decomposition of a time function into exponentials, $e^{j\omega t}$ which exist for all time. We should consider the signal that is processed with the DFT calculations as being extended periodically for all time as discussed in the previous section. That is, the data window represents one period of a periodic signal. The sampling rate and the length of the data window determine the frequency resolution of the calculations. The wavelet transform introduces an alternative to these limitations.

A windowed Fourier transform can be written as in Equation (4.44)

$$X(\omega, t) = \int_{-\infty}^{\infty} x(\tau)h(t - \tau)e^{-j\omega\tau}\, d\tau \qquad (4.44)$$

while the wavelet transform takes the form of Equation (4.45)

$$X(s, t) = \int_{-\infty}^{\infty} x(\tau)\left[\frac{1}{\sqrt{s}}\psi\left(\frac{\tau - t}{s}\right)\right] d\tau \qquad (4.45)$$

where s is a scale parameter.
If h(t) has Fourier transform H(ω) then h(t/s) has Fourier transform H(sω). Note that for a fixed h(t), that a large value of s compresses the transform while small

value of s spreads the transform in frequency. There are a few requirements on a signal h(t) to be the 'mother wavelet' (essentially that h(t) have finite energy and be a band-pass signal).

Wavelets are a family of functions obtained from the 'mother wavelet', ψ, by dilating and translating. The discrete version is given by Equation (4.46).

$$DWT(x, m, n) = \frac{1}{\sqrt{a^m}} \cdot \sum_k x(k) \psi \left(\frac{k - na^m}{a^m} \right) \tag{4.46}$$

$$\psi_{mn}(t) = a^{-\frac{m}{2}} \cdot \psi \left(\frac{t - na^m}{a^m} \right) \tag{4.47}$$

Let $\psi(t)$ be a rectangular pulse with one second duration as shown in Figure 4.13. The functions $\psi_{0n}(t) = \psi(t - n)$ are shifted (translated) replicas of the one second pulse. The functions $\psi_{1n}(t)$ in Equation (4.47), with $a = 2$, are dilated (have a two second duration) as shown in Figure 4.13.

$$\psi_{1n}(t) = \frac{\sqrt{2}}{2} \psi \left(\frac{1}{2}(t - 2n) \right) \tag{4.48}$$

Note that the set of functions $\{\psi_{0n}(t)\}$ are orthogonal as are the set $\{\psi_{1n}(t)\}$, Th is property is maintained as m increases. The functions $\{\psi_{mn}(t)\}$ have a 2m second duration and are orthogonal to each other. If we denote the space spanned by $\{\psi_{mn}(t)\}$ as V_m then $V_0 \supset V_1 \supset V_{20} \cdots$, i.e. V_{m+1} is coarser than V_m. There are a number of common mother wavelets both continuous and discrete. The coarseness mentioned for the rectangular wavelets translates into different frequency resolution as m changes. Smaller m gives higher frequency resolution. Some rectangular wavelets and their Fourier transforms are shown in Figure 4.14. Discrete wavelets include: Daubechies wavelets, Haar wavelets, Mathieu wavelets, Legendre wavelets and Villasenor wavelet.

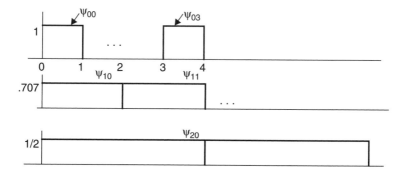

Figure 4.13 Rectangular wavelets

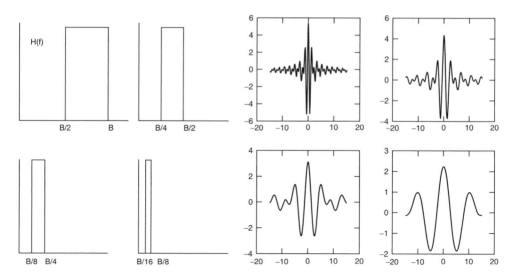

Figure 4.14 Rectangular wavelets and their Fourier transforms

Wavelets are also related to filter banks. To illustrate this, consider using wavelets to compress a discrete waveform. Figure 4.15 shows a filter bank in which the incoming signal h(k) is split into a high-pass and a low-pass signal at the far left of the figure. Each half is samples at half the original sampling rate, the half band high-pass signal is sent forward as $h_1(k_1)$. and the low-pass signal is split again. Consider the sequence $x(n) = [-2 \ -28 \ -46 \ -44 \ -20 \ 12 \ 32 \ 30]$ with the low-pass filter being the average of two consecutive signals $(x(n) + x(n+1))/2$ and the high-pass filter being the difference $(x(n) - x(n+1))/2$.
The resulting signals shown in Figure 4.15 are

$$h_1(k_1) = [13 - 1 - 16 \ 1], h_2(k_2) = [7 - 8.5]$$

$$h_3(k_3) = [7.75], l_3(k_3) = [-0.75]$$

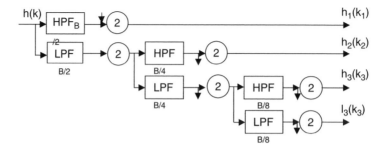

Figure 4.15 Filter bank

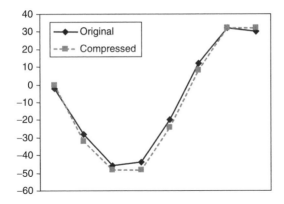

Figure 4.16 The original and reconstructed compressed waveforms

If we truncate to three levels 0, 8, and 16 (compressing the signal)

$$h_1(k_1) = [16\ 0 - 16\ 0], h_2(k_2) = [8 - 8]$$
$$h_3(k_3) = [8], l_3(k_3) = [0]$$

and reconstruct the original sequence the original and compressed signal are shown in Figure 4.16.

4.10 Elements of artificial intelligence

Suggested application of a limited number of AI concepts in protection has emerged in the literature. It is attractive to imagine some powerful new technique solving existing protection problems. Two techniques, Artificial Neural Networks, and decision trees, are based on training the relay with simulation and even field data to differentiate between fault and no fault conditions. The third involves the use of autonomous software agents distributed through the protection system.

4.10.1 Artificial neural networks

Artificial Neural Networks (ANNs) had their beginning in the 'perceptron' which was designed to recognize patterns.[6] In some sense ANNs are modeled after the structure of the brain.[7] The number of papers suggesting relay application has soared. The attraction is the use of ANNs as pattern recognition devices that can be trained with data to recognize faults or inrush or other protection effects. The basic feed forward neural net is composed of layers of neurons as shown in Figure 4.17.

The function F is either a threshold function or a saturating function such as a symmetric sigmoid function.(ψ in Figure 4.17) The weights w_i are determined by

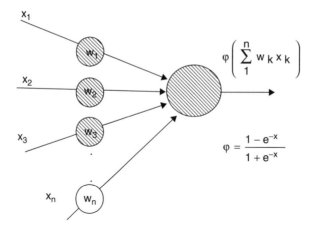

Figure 4.17   Artificial Neural Networks

training the network. (so called back propagation). The training process is the most difficult part of the ANN process. Typically simulation data such as that obtained from EMTP is used to train the ANN. A set of cases to be executed must be identified along with a proposed structure for the net. The cases include the data the relay would see and a declaration that the data belongs to fault case or a non-fault case. The Matlab Neural network toolbox can perform the training but only after a structure is chosen. A typical structure is shown in Figure 4.17.

The structure required is described in terms of the number if inputs, the number of hidden layers, and the number of neurons in the various layers, and the output(s). An example might be a net with 12 inputs, and a 4, 3, 1 layer structure. There would be 4×12 plus 4×3 plus 3×1 or 63 weights to be determined. Clearly a lot more than 63 training cases are needed to learn 63 weights. In addition some cases not used for training are needed for testing.

Once the weights are learned the designer is frequently asked how the ANN will perform when some combination of inputs are presented to it. The ability to answer such questions is very much a function of the breadth of the training sequence. The proposed protective relaying applications of ANNs include:

- high impedance fault detection
- transformer protection
- fault classification
- fault direction determination
- fault location
- adaptive reclosing
- rotating machinery protection.

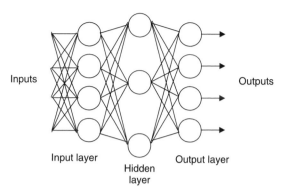

Inputs   Outputs

Input layer   Output layer
                Hidden
                layer

Figure 4.18 Layers

4.10.2 Decision trees

Decision trees are commonly used in data mining and have the advantage of simplicity compared to neural networks. A classification tree is shown in Figure 4.18. The tree represents a hypothetical experiment formed by the results from 100 transient stability simulations. Each of the simulations is declared stable or unstable and a set of prefault measurements (angles, voltages, and flows) are attached to that result. There are 50 stable cases and 50 unstable cases and each set of measurements include 80 measurements. The tree is formed by processing a data set of 100×81 numbers.

$$(m_{i1}, m_{i2}, m_{i3}, \cdots, m_{i80}, y_i) \ i = 1, 2, 3, \cdots, 100 \tag{4.49}$$

The y_i are 1 or 0 while the m_i are angles or voltages or real or reactive power flows. The white boxes in Figure 4.19 are splitting nodes where the value of a measurement is used to branch to the right or the left. The top box uses some measurement, say, m_{ik1} to divide the 100 cases into 55 on the right and 45 to the left.

$$\text{left} \quad \text{if} \quad m_{ik1} < N1$$

$$\text{right} \quad \text{if} \quad m_{ik1} > N1 \tag{4.50}$$

Note the 55 to the right are not all stable or unstable, nor are 45 to the left. Also observe the branching is on a single measurement. There are five additional splitting nodes, with the number going left and right labeled for each. The shaded nodes are the terminal nodes and for the example are correct. In large real problems, there may be a small number of mis-classifications permitted. The example has far fewer training cases than normally would be used. The actual tree structure resulted from a training set of 4150 simulation cases which was tested on an additional 4385 cases with a total error rate less than 1%.

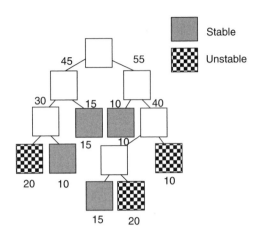

Figure 4.19   A classification tree

Note that the tree has selected six variables out of the 80 measurements to complete the tree. This is frequently the most important feature of the tree. One can hypothesize that a large number of PMUs are available for the measurements and the decision tree training algorithm will select the best location for the placement of PMUs out of the original offering. This is not the case for the ANN. If we imagined 80 inputs for the ANN we could hope some of the weights were small enough to be omitted but it can not be guaranteed. In addition the size of neural net with 80 inputs would make training difficult.

Experience with both ANNs and decision trees is that they both interpolate well but have difficulty extrapolating. That is, if we imagine the set of measurements that are used to train the ANN or classification tree as a region in measurement space then a new set of measurements (not used in training) that lies within the training set is unlikely to cause an error in classification. A new point that is outside the training set, in any sense, has a larger chance of being mis-classified. The errors tend to occur at the boundaries.[8] This means that a great deal of engineering experience is required in selecting the training cases.

4.10.3   Agents

An agent is a self-contained piece of software that has the properties of autonomy and interaction. Agents in relays would be capable of autonomously interacting with each other. It has been proposed that using geographically distributed agents located in every relay can improve on more traditional isolated component system. A suggested research area is the use of agent-based backup protection system. The fact that agent relays would be capable of autonomously interacting with each other could make a distributed systems work effectively. This flexibility and autonomy adds reliability to the protection system because any given agent-based relay can continue to work properly despite failures in other parts of the protection system.

It is certainly necessary to explore the expected communication traffic patterns in order to make agents more intelligent and robust towards network conditions. The object models of multifunctional distance relays in IEC 61850 may provide an environment for agents. Message structure and strategies for such systems have been studied in simulation.[9–12]

4.11 Conclusion

This chapter has continued in the vein of previous chapter with more of the mathematical basis for computer relaying. In this chapter there is more emphasis on digital signal processing. Discrete time signals and systems, the z transform, and discrete time Fourier transforms have parallels in the material in Chapter 3, while some of the material in the Section 4.6 through Section 4.10 is unlike Chapter 3. The concept of windowing and the behavior of various windows is less of an issue in relaying then in other uses of the measurements that relays make. Synchronized phasor measurements in Chapter 8 are such an application. The windowing material is then the background for some of the signal processing involved in processing synchronized phasor measurements. Wavelets are, in a sense, a generalization of Fourier analysis and have found wide application in other fields. Artificial intelligence, including ANNs, decision trees, and agents are concepts that have been suggested in the literature to improve relaying. The first two involve training the relay with simulation or field data while the third allows autonomous behavior of software installed in relays. Both concepts have encountered the kind of resistance encountered by the first attempts at digital relaying. The relay engineer is reluctant to accept the possibility of a relay, at some time in the future, behaving in a manner that cannot be explained.

Problems

4.1 Given a causal LTI system with input y(t) and output x(t) described by

$$X(s) = \frac{Y(s)}{s^2 + 3s + 2}$$

Write a difference Equation for a discrete time system which is impulse invariant (with a sampling time of T seconds) to the continuous system. i.e.,

$$h[n] = h_a(nT)$$

4.2 Find a discrete tine sequence x[n]; $-\infty < n, < \infty$ which has z transform

$$X(z) = \frac{1}{z^2(z - 0.5)(z + 1)}; \quad 0.5 < |z| < 1$$

4.3 Determine, in each case, whether $x[n]$ has a discrete-time Fourier transform, $X(z)$, and if so find it.

(a) $x[n] = 2^n u[n]$
(b) $x[n] = 2^n u[-n]$
(c) $x[n] = 2^{-n} u[n]$
(d) $x[n] = 2^{-n} u[n] + 3^n u[-n-1]$
(e) $x[n] = x_1[n] * x_2[n] \quad x_1[n] = \left(\frac{2}{3}\right)^n u[n] \quad x_2[n] = e^{-3n} u[n]$

4.4 A discrete time system is used to model a causal continuous time system modeled by

$$\frac{dx(t)}{dt} + 0.9x(t) = y(t)$$

By replacing $\frac{dx(t)}{dt}$ with the forward difference approximation

$$\frac{x((n+1)T) - x(nT)}{T}$$

(a) Sketch the magnitude of the magnitude of the frequency response of the discrete time system.
(b) For $T = 10/9$ sketch the magnitude of the frequency response of the discrete time system.
(c) For what values of T is the discrete time system unstable?

4.5 Find the transfer function for two discrete time systems which simulate the system $H(s) = \frac{50}{(S+W_c)^3}$ both an impulse invariant: system and one obtained with the bilinear transformation.

4.6 If the impulse response of a discrete time system is given by

$$h[n] = \left(\frac{1}{2}\right)^n u[n]$$

Find the output $x[n]$ if the input is $y[n] = u[n]$. Use convolution.

4.7 If it exists find the z transform (the function and the ROC) for

(a) $x[n] = u[n] \left(\frac{1}{3}\right)^n + u[-n-1]3^n$
(b) $x[n] = u[n](3)^n + u[-n-1]\frac{1}{3^n}$
(c) $x[n] = 1; \quad 0 \le n \le (N-1)$

    $x[n] = 0 \quad$ elsewhere

(d) $x[n] = n \left(\frac{1}{2}\right)^n u[n+1]$

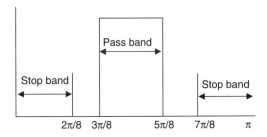

Figure P4.20 **Figure** for Problem 4.10.

4.8 Find x[n] where X(z) is as indicated

(a) $X(z) = \frac{z}{z^2 - .5z + .06};\quad 0.3 < |z| < \infty$

(b) $X(z) = \frac{4}{z3(4z-1)};\quad 0 < |z| < 1/4$

(c) $X(z) = \frac{z^2}{(z-1)(z+2)};\quad 1 < |z| < 2$

4.9 In each case find the response to an input y[n] = u[n]

(a) $x[n] - 2x[n-1] + x[n-2] = y[n]$

(b) $x[n] - x[n-2] = y[n] - y[n-1]$

4.10 Use the Matlab function firpm.m to design a 64th order band-pass filter with the desired pass band and stop band shown.

References

[1] http://zone.ni.com/devzone/cda/tut/p/id/4844

[2] Oppenheim, A. V., and R. W. Schafer (1989) *Discrete-Time Signal Processing*, Prentice-Hall, pp. 447–448.

[3] Harris, F. J. (1978) On the use of windows for harmonic analysis with the discrete Fourier transform. Proceedings of the IEEE, vol. 66, pp. 51–84.

[4] D'Antona, G. and Ferrero, A. (2006) *Digital Signal Processing for Measurement Systems*, New York: Springer Media, Inc., pp. 70–72.

[5] Rabiner, L. R., McClellan, J. H. and Parks, T. W. (1975) FIR digital filter design techniques using weighted Chebyshev approximations, Proceedings of the IEEE, vol. 63, pp. 595–610.

[6] Rosenblatt, F. The Perceptron: A probabilistic model for information storage and organization in the brain, Cornell Aeronautical Laboratory, *Psychological Review*, vol. 65, no. 6, pp. 386–408.

[7] Haykin, S. (1994) *Neural Networks: A Comprehensive Foundation*, Macmillian.

[8] Sheng, Y. and Rovnyak, S. M. (2002) Decision trees and wavelet analysis for power transformer protection, IEEE Trans. on Power Delivery, vol. 17, no. 2.

[9] IEC 61850-7-2 (2003–2005) Communication networks and systems in substations, Part 7-2: Basic communication structure for substations and feeder equipment – Abstract communication service interface (ACSI).

[10] Hopkinson, Wang, X., Giovanini, R., Thorp, J. S. et al. (2006) EPOCHS A platform for agent-based electric power and communication simulation built from commercial off-the-shelf components, IEEE Trans. on Power Systems, vol. 21, no. 2, pp. 548–558.

[11] Apostolov, A. P. (2001) Object models of multifunctional distance relays, Proceedings of the Power Engineering Society Summer Meeting, pp. 1157–1162.

[12] Apostolov, A. P. (2001) Multi-agent systems and IEC 61850, Proceedings of the Power Engineering Society General Meeting, 2001, pp. 1–6.

5

Transmission line relaying

5.1 Introduction

Historically, line relaying algorithms represent most of the early activity in computer relaying. Researchers perceived that line relaying offered the greatest challenge, along with the greatest possibility for improved performance. In this chapter we will examine many of these algorithms and attempt to draw some general conclusions about the characteristics of several types of algorithms.

A number of algorithms can be regarded as impedance calculations in that the fundamental frequency components of both voltages and currents are obtained from the samples. The ratios of appropriate voltages and currents then provide the impedance to the fault. The performance of all of these algorithms is dependent on obtaining accurate estimates of the fundamental frequency components of a signal from a few samples. Within this class of algorithm both Fourier and curve fitting techniques are used to estimate the fundamental frequency components. If the signal in question were a pure sinusoid then virtually every algorithm ever suggested would work perfectly. The distinction between algorithms of this type is in their behavior when signals other than the fundamental frequency are present in the voltage and current.

Another type of algorithm is based on a series R-L model of the transmission line. Rather than using a single frequency model, such as the impedance, this approach has the apparent advantage of allowing all signals that satisfy the differential equation to be used in estimating the R and L of the model.

Line relaying algorithms have been compared under the assumption that the currents and voltages were composed of the fundamental and certain combinations of harmonics.[1] It is our contention that, given all the situations in which a line relay must operate – i.e. a changing system configuration, variable fault incidence angle etc. – the non-fundamental frequency components seen by the relay must be

Computer Relaying for Power Systems 2e   by A. G. Phadke and J. S. Thorp
© 2009 John Wiley & Sons, Ltd

regarded as a random process. The nature of this random process is then a primary issue in evaluating relaying algorithm performance.

In the following sections the sources of error in the signals will be investigated and a statistical description of the errors will be postulated. For a large class of algorithm any non-fundamental frequency signal constitutes an error. The connection between the error model and the algorithm used to estimate the parameters of interest will be established. Before proceeding, certain notation and concepts common to all algorithms will be illustrated for a simple example. The following notation will be used in discussing all of the algorithms:

$y(t) = $ The instantaneous value of an AC waveform, a voltage or a current
$y_k = $ The kth sample value of $y(t)$
$\omega_o = $ The fundamental power system frequency in radians per second
$\Delta t = $ The fixed interval between samples, i.e.
$y_k = y(k\Delta t)$
$\theta = $ The fundamental frequency angle between samples, i.e. $\theta = \omega_o \Delta t$

To illustrate some of the common features of algorithms based on a waveform model suppose $y(t)$ is assumed to be of the form

$$y(t) = Y_c \cos \omega_o t + Y_s \sin \omega_o t \tag{5.1}$$

where Y_c and Y_s are real numbers, Further, assume samples are taken at $-\Delta t$, 0, and Δt

$$y_{-1} = y(-\Delta t)$$

$$y_o = y(0)$$

$$y_1 = y(\Delta t) \tag{5.2}$$

The samples are related to the amplitudes Y_c and Y_s through

$$\begin{bmatrix} y_{-1} \\ y_o \\ y_1 \end{bmatrix} = \begin{bmatrix} \cos\theta & -\sin\theta \\ 1 & 0 \\ \cos\theta & \sin\theta \end{bmatrix} \begin{bmatrix} Y_c \\ Y_s \end{bmatrix} \tag{5.3}$$

where θ is the fundamental frequency angle between samples. It is clear that two samples are sufficient to determine Y_c and Y_s if the signal is described by Equation (5.1). For example, $Y_c = y_0$ and $Y_s = (y_1 - y_0 \cos\theta)/\sin\theta$ satisfy the last two equations in (5.3). The use of three samples is an attempt to provide some immunity to additional terms (harmonics or random terms) in Equation (5.1). As such, a least squares solution to Equation (5.3) would seem appropriate. Using

Equation (3.75) the least square solution is

$$\begin{bmatrix} \hat{Y}_c \\ \hat{Y}_s \end{bmatrix} = \begin{bmatrix} 1 + 2\cos^2\theta & 0 \\ 0 & 2\sin^2\theta \end{bmatrix}^{-1} \begin{bmatrix} \cos\theta & 1 & \cos\theta \\ -\sin\theta & 0 & \sin\theta \end{bmatrix} \begin{bmatrix} y_{-1} \\ y_o \\ y_1 \end{bmatrix}$$

$$\hat{Y}_c = \frac{[y_1 \cos\theta + y_o + y_{-1} \cos\theta]}{1 + 2\cos^2\theta} \tag{5.4}$$

$$\hat{Y}_s = \frac{[y_1 - y_{-1}]}{2\sin\theta} \tag{5.5}$$

A more general solution to Equation (5.3) (not a least square solution) is in the form

$$Y_c = \hat{Y}_c + c_1[y_1 - 2y_o \cos\theta + y_{-1}] \tag{5.6}$$

$$Y_s = \hat{Y}_s + c_2[y_1 - 2y_o \cos\theta + y_{-1}] \tag{5.7}$$

where c_1 and c_2 are arbitrary constants. The bracketed terms in Equations (5.6) and (5.7) are zero if the signal is described by Equation (5.1). Two of the early algorithms correspond to particular choices of c_1 with $c_2 = 0$. The Mann-Morrison algorithm[2] corresponds to $c_2 = 0$ and

$$c_1 = \frac{-\cos\theta}{1 + 2\cos^2\theta}$$

while the Prodar 70 algorithm[3] corresponds to $c_2 = 0$ and

$$c_1 = \left[\frac{1}{\sin^2\theta} - \frac{\cos\theta}{1 + 2\cos^2\theta} \right]$$

The original versions of both algorithms also are based on the assumption that θ is small enough so that the small angle assumptions that, $\cos(\theta) \approx 1$ and $\sin(\theta) \approx 0$ are appropriate and represent approximations of derivative terms from samples.

In order to examine some general properties of all such algorithms let us take the version with c_1 and c_2 equal to zero and examine the computation as time passes and more samples become available. An algorithm based on the last three samples y_{k-1}, y_k, and y_{k+1}, would have the form

$$\hat{Y}_c^{(k)} = \frac{[y_{k+1} \cos\theta + y_k + y_{k-1} \cos\theta]}{1 + 2\cos^2\theta} \tag{5.8}$$

$$\hat{Y}_s^{(k)} = \frac{[y_{k+1} - y_{k-1}]}{2 \sin \theta} \tag{5.9}$$

where the superscript k indicates calculations centered at the kth sample. If $y(t)$ were a pure sinusoid as in Equation (5.1) then (in fact, for any choice of c_1 and c_2)

$$Y_c^{(k)} = Y_c \cos k \, \theta + Y_s \sin k \, \theta \tag{5.10}$$

$$Y_s^{(k)} = Y_s \cos k \, \theta - Y_c \sin k \, \theta \tag{5.11}$$

In polar form

$$\left| Y^{(k)} \right| = \sqrt{(Y_c^{(k)})^2 + (Y_s^{(k)})^2} \tag{5.12}$$

$$\varphi^{(k)} = \tan^{-1} \left[\frac{Y_s^{(k)}}{Y_c^{(k)}} \right] = \tan^{-1} \left[\frac{Y_s}{Y_c} \right] - k\theta \tag{5.13}$$

It can be seen from Equation (5.13) that the computed phasor has the correct amplitude but rotates i.e. the angle $\varphi^{(k)}$ decreases by the angle θ at each sample point Depending on the application, it may be necessary to correct for the rotation. If the ratio of voltage and current phasors is to be used for an impedance calculation, however, then the rotation will cancel in the division.

The algorithm described by Equations (5.8) and (5.9) has a *data window* of three samples, that is, as a new sample becomes available, the oldest of the three sample values is discarded and the new sample value is included in the calculation. Each sample is then used in three calculations, once as y_{k+1}, once as y_k, and once as y_{k-1}. The calculation in Equations (5.8) and (5.9) must then be completed by the microprocessor before the next sample is produced. In practice a great deal more must actually be computed as we will see later.

A moving data window of three samples is shown in Figure 5.1 for an ideal voltage waveform sampled at 12 samples per cycle. The voltage decreases instantaneously at the fault instant. The window labeled W1 contains three samples of pre-fault data, the windows W2 and W3 contain both pre- and post-fault data, and the window W4 has only post-fault data. The calculations in Equations (5.8) and (5.9) will produce the correct phasors in the windows containing pure pre-fault or post-fault samples. The data in the windows W2 and W3, however, cannot be fitted to a pure sinusoid and the computed phasor is of little meaning. It can be verified, however, that the computed phasor does not fit the three samples. It should be noted that a two sample window will always fit the data although the fit to one pre-fault and one post-fault sample is equally meaningless.

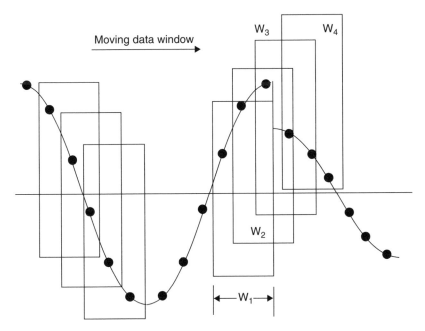

Figure 5.1 Moving three sample data window on a voltage waveform. W1 is pre-fault, W2 and W3 contain both pre-fault and post-fault samples, and W4 is pure post-fault

There are several conclusions that can be drawn from the simple three sample algorithm and the Figure. The sample time, Δt, determines the amount of time the microprocessor has to complete its calculations. The example with 12 samples per cycle has $\Delta t = 1.3889$ msec for a 60 Hz system. On a 50 Hz system 20 samples a cycle yields $\Delta t = 1$ msec. Existing algorithms use sampling rates from four to 64 samples a cycle. Clearly high sampling rates require more powerful processors or particularly simple algorithms.

The second issue is that of the length of the data window. Recognizing that the results obtained when the window contains both pre-fault and post-fault samples are unreliable, it seems reasonable to wait until the results are reliable (when the window contains only post-fault data) before making relaying decisions. It is important that a technique that senses this transition region be developed. The response of the example algorithm in this transition region is a function of the parameters c_1 and c_2 in Equations (5.6) and (5.7). Since a longer window takes longer to pass over the fault instant, it is clear that faster decisions can be made by short window algorithms. Unfortunately, as we will see in Section 5.5, the ability of an algorithm to reject non-fundamental frequency signals is a function of the length of the data window. In other words, there is an inherent inverse relationship between relaying speed and accuracy. While the algorithm represented by Equations (5.8) and (5.9) yields the correct phasor if the signal y(t) is given by Equation (5.1), we must recognize that the signal to be sampled is more accurately given by:

$$y(t) = Y_c \cos\omega_o t + Y_s \sin\omega_o t + \varepsilon(t) \qquad (5.14)$$

It is the nature of the signal $\varepsilon(t)$ in Equation (5.14) that must be understood in order to evaluate line relay performance.

5.2   Sources of error

The post-fault current and voltage waveforms fail to be pure fundamental frequency sinusoids for a variety of reasons. The most predictable non-fundamental frequency term is the decaying exponential which can be present in the current waveform. For the series R-L model of the line shown in Figure 5.2, assuming zero pre-fault current and a steady state fault current of the form $I \cos(\omega_0 t - \varphi)$, the instantaneous current for a fault at time t_0 is given by

$$i(t) = I \cos(\omega_0 t - \varphi) - [I \cos(\omega_0 t_0 - \varphi)]e^{-(t-t_0)R/L} \qquad (5.15)$$

The second term in Equation (5.15) decays exponentially with the time constant of the line. This term is the main cause of transient over-reach in high speed relays, and must be eliminated if computer relaying at fractional cycle speed is to be achieved. For a typical EHV line the time constant is in the range of 30–50 ms. The initial amplitude of the exponential component can be as large as the peak of the fault current, as shown in Figure 5.3. The situation can be even more complicated near a large generator. The exponential term is not an error to algorithms based on a differential equation description of the line, since the exponential satisfies the differential equation. If the time constant of the line is known, then the decay can be removed with an external filter or even in software (for algorithms which see the exponential decay as an error). The dependence of the time constant on fault resistance makes the removal less effective for high resistance faults.

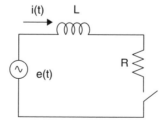

Figure 5.2   Series R-L model of a transmission line

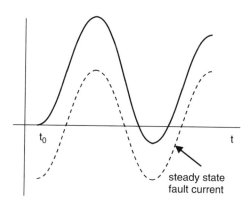

Figure 5.3   Fault current. The dashed curve is the steady state fault current

Other non-fundamental frequency terms are not so easy to remove because they are not so easy to predict. The current and voltage transducers contribute some of these signals. For example, as shown in Section 2.6, a capacitive coupled voltage transformer has a transient response to the abrupt change in voltage shown in Figure 5.1. High frequency signals associated with the reflection of waveforms between the bus and fault may be present, as will be developed in Chapter 8. The nonlinear behavior of the fault arc may produce harmonic signals. In addition, as shown in Section 1.5, the A/D converter contributes errors due the least significant bit in the conversion and due to timing errors, i.e. the samples are not exactly Δt sec appart. Most of these signals have considerable high frequency content and can be reduced through the use of an anti-aliasing filter.

Since the waveform is being sampled at a rate of $f_s = 1/\Delta t$ Hz the Nyquist sampling theorem from Section 3.5 implies that the signal should be filtered with a filter having a cut-off frequency of $f_s/2$ to avoid aliasing. As mentioned in Chapter 1, such a filter will remove the high frequency error signals described above but will contribute a transient response of its own; also, drift over time of the component values in such filters (particularly active realizations of such filters) are sources of error.

Finally, the power system itself is a source of non-fundamental frequency signals. Consider the single phase model of three lines and two generators shown in Figure 5.4. It is assumed that the lines are identical but one source is strong and one is weak. The lines are assumed to be 100 miles of typical 765 kV line. If a fault is applied at 60% of the protected line, the voltage seen by the relay is shown in Figure 5.5. The smooth curve is the voltage that would result if the capacitors were removed from Figure 5.5. It can be seen that the inclusion of the capacitors has produced at least two non-fundamental frequency signals. These non-fundamental frequencies are natural frequencies of the system which are excited by the application of the fault. Since the network is fixed if the fault location is

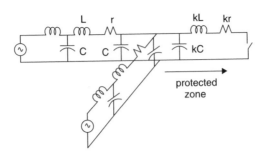

Figure 5.4 Single-phase power system model

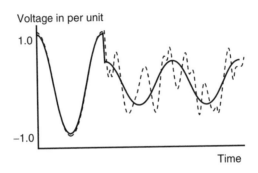

Figure 5.5 Voltage waveforms for a fault at 60% of the line length

held constant, it follows that the natural frequencies are determined by the fault location.

Figure 5.6 shows a family of voltage waveforms produced by altering the fault incidence angle. It can be seen that the phase of the non-fundamental frequency components is a function of the fault incidence angle. As the fault location is changed as shown in Figure 5.7, the frequency of the non-fundamental frequency components

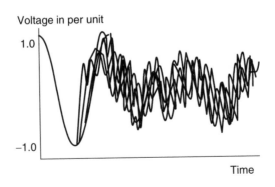

Figure 5.6 Family of voltage waveforms for faults at 60% of the line length

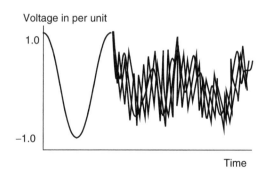

Figure 5.7   Family of voltage waveforms for varying fault locations

changes. A similar effect can be produced by altering the network structure behind the fault. Experiments on a model power system combining the effect of changing the structure of the network feeding the fault and the fault type and location have been reported.[4] The conclusion is that an important part of the non-fundamental frequency signal $\varepsilon(t)$ in Equation (5.14), at least for high voltage lines, is due to the network itself. These signals depend on the fault location and on the nature of the system feeding the fault and, as such, are not predictable. Example 3.20 provides a model of such a process. If we model the fault incidence angle as the random phase and the fault location and network structure as the mechanism producing the random frequency then we can think of the power spectrum of Example 3.20 as the power spectrum of the signal in $\varepsilon(t)$ in Equation (5.14). It should be recognized that each realization of such a process is a rather deterministic looking signal, such as that shown in Figure 5.5. (This is counter intuitive if one expects a realization of a random process to look noisy.) The randomness is present because, considering the ensemble of times that the relay is expected to operate, the frequency and phase of the signal cannot be predicted.

Considering the $\varepsilon(t)$ in Equation (5.14) as a random process, it is reasonable to consider the anti-aliasing filter and the algorithm taken together as filtering the random process, as in Section 3.8. The frequency response of the algorithm is then an important part of the filtering process. To obtain the frequency response of the algorithm we should compute the response of the algorithm (the phasor for Equations (5.6) and (5.7), for example) when the input signal is of the form $e^{j\omega t}$.

Example 5.1

If $y(t) = e^{j\omega t}$

$$y_{-1} = e^{-j\omega \Delta t}$$

$$y_0 = 1$$

$$y_1 = e^{j\omega \Delta t}$$

and Equations (5.4) and (5.5) yield

$$\hat{Y}_c = \frac{e^{j\omega\Delta t}\cos\theta + 1 + e^{-j\omega\Delta t}\cos\theta}{1 + 2\cos^2\theta}$$

$$\hat{Y}_c = \frac{1 + 2\cos\theta\cos\omega\Delta t}{1 + 2\cos^2\theta}$$

and

$$\hat{Y}_s = \frac{e^{j\omega\Delta t} - e^{-j\omega\Delta t}}{2\sin\theta}$$

$$\hat{Y}_s = j\frac{\sin\omega\Delta t}{\sin\theta}$$

It can be seen, for example, that, if $y(t) = \text{Re}\{e^{j\omega_0 t}\} = \cos\omega_0 t$, $\hat{Y}_c = 1$, and $\hat{Y}_s = 0$. In general, if $y(t) = \text{Re}\{e^{j(\omega t + \varphi)}\} = \cos(\omega t + \varphi)$ then

$$\hat{Y}_c = \cos\varphi\left[\frac{1 + 2\cos\theta\cos\omega\Delta t}{1 + 2\cos^2\theta}\right]$$

$$\hat{Y}_s = \sin\varphi\left[\frac{\sin\omega\Delta t}{\sin\theta}\right]$$

The two bracketed terms are plotted in Figure 5.8 along with $\sqrt{(\hat{Y}_c^2 + \hat{Y}_s^2)/2}$ (the magnitude for $\varphi = 45°$) for $\theta = 30°$ (12 samples per cycle). Since two quantities are being computed, there are two frequency responses. The choice of angle φ that is used in presenting the magnitude is somewhat arbitrary. We will use $\varphi = 45°$ for consistency.

Different frequency responses are obtained for different values of c_1 and c_2 in Equations (5.6) and (5.7). The Prodar 70 algorithm was specifically designed to

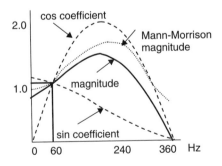

Figure 5.8 Frequency responses of Equations (5.4) and (5.5) along with the Mann-Morrison algorithm

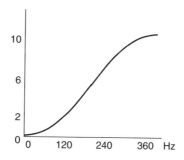

Figure 5.9   Frequency response of the Prodar 70 algorithm

reduce low frequency response. The frequency responses of the Mann-Morrison and the Prodar 70 algorithms at the same sampling rate of 12 samples per cycle are shown in Figure 5.8 and Figure 5.9.

The algorithms compared in Figures 5.8 and 5.9 represent three samples at a rate of 12 samples per cycle, or a quarter of a cycle. As such, they do not reject non-fundamental frequency components (particularly the third harmonic) as well as might be necessary in many applications. These algorithms are fast in that the short window will be entirely in the post-fault region quickly. We will see later that longer window algorithms have a greater ability to reject non-fundamental frequencies at the expense of a longer decision time (the longer window takes longer to clear the instant of fault inception). This is a general result which we will develop throughout this chapter. That is, there is an inherent speed-reach limitation in line relaying caused by the presence of random signals in the measured voltages and currents. The reach of the relay (the setting of the boundary of the zone of protection) is directly related to the accuracy of the estimates formed by the algorithm.

5.3   Relaying as parameter estimation

If Equation (5.14) is generalized to include a number of known signals including the fundamental frequency components, a general problem in estimating the coefficients of the known signals can be posed. The resulting solution encompasses a number of relaying algorithms. The signal to be sampled is written as

$$y(t) = \sum_{n=1}^{N} Y_n s_n(t) + \varepsilon(t) \tag{5.16}$$

or

$$y_k = \sum_{n=1}^{N} Y_n s_n(k\Delta t) + \varepsilon(t) \tag{5.17}$$

where the signals s(t) are assumed known but the coefficients Y_n are unknown. The obvious choices for the signals include:

$$\left. \begin{array}{l} s_1(t) = \cos \omega_0 t \\ s_2(t) = \sin \omega_0 t \end{array} \right\} \text{ the previous example}$$

$$\left. \begin{array}{l} s_3(t) = \cos 2\omega_0 t \\ s_4(t) = \sin 2\omega_0 t \end{array} \right\} \text{ the second harmonic}$$

$$\left. \begin{array}{l} \vdots \end{array} \right\} \text{ other harmonics}$$

$s_N(t) = e^{-(R/L)t}$ the exponential offset

The problem then is to estimate the coefficients Y_n from the measurements Y_k. The least squares technique from Chapter 3 is appropriate if we write

$$\begin{bmatrix} y_1 \\ y_2 \\ \vdots \\ y_k \end{bmatrix} = \begin{bmatrix} s_1(\Delta t)s_2(\Delta t) \cdots s_N(\Delta t) \\ s_1(2\Delta t)s_2(2\Delta t) \cdots s_N(2\Delta t) \\ \vdots \\ s_1(k\Delta t)s_2(k\Delta t) \cdots s_N(k\Delta t) \end{bmatrix} \begin{bmatrix} Y_1 \\ Y_2 \\ \vdots \\ Y_N \end{bmatrix} + \begin{bmatrix} \varepsilon_1 \\ \varepsilon_2 \\ \vdots \\ \varepsilon_k \end{bmatrix}$$

or

$$y = S\,Y + \varepsilon \tag{5.18}$$

where Equation (5.18) represents k equations in N unknowns. It is clear that $k \geq N$ is required in order to estimate all of the N parameters. If the error vector ε is assumed to have zero mean, i.e. $E\{\varepsilon\} = 0$ and a covariance matrix

$$E\{\varepsilon\varepsilon^T\} = W \tag{5.19}$$

then the solution from Section 3.7 (Equation (3.77)) is appropriate and yields

$$\hat{Y} = (S^T W^{-1} S)^{-1} S^T W^{-1} y \tag{5.20}$$

The estimate given by Equation (5.20) is unbiased, i.e.

$$E\{\hat{Y}\} = E\{Y\} \tag{5.21}$$

and has a covariance

$$E\{(\hat{Y} - Y)(\hat{Y} - Y)^T\} = (S^T W^{-1} S)^{-1} \tag{5.22}$$

5.3.1 Curve fitting algorithms

The algorithms developed along the lines of Equation (5.20)[5,6] have assumed, in essence, that **W** is a multiple of a unit matrix (the errors are uncorrelated and independent from sample to sample and have a constant covariance) so that the least squares solution is

$$\hat{Y} = (S^T S)^{-1} S^T y \tag{5.23}$$

The matrix $(S^T S)^{-1} S^T$ can be computed off-line and stored. In fact, only the two rows of the matrix that correspond to $\cos \omega_0 t$ and $\sin \omega_0 t$ are needed for impedance relaying (the ratio of the fundamental frequency voltage and current is sufficient). If the DC offset (the exponential decay in the fault current) is included, the matrix $(S^T S)^{-1} S^T$ is full, so that a full set of 2k numbers is needed. In addition to a number of harmonics and the magnitude of the offset, the time constant of the line itself has been included in the parameters to be estimated.[5]

5.3.2 Fourier algorithms

If, as in the curve fitting algorithms, **W** is assumed to be a multiple of a unit matrix and if the DC offset is removed with an analog filter or with a separate piece of software (see Section 5.3.8) then Equations (5.20) or (5.23) becomes particularly simple. If only the fundamental and harmonics are included in the signal set $\{s_n(t)\}$ and an even number of sample panning a full period is used then Equation (5.23) becomes a rectangular form of the DFT. With K samples per cycle (K/2-1) harmonics can be computed. Using the orthogonality of the sine and cosine terms, the ij^{th} entry of the matrix $S^T S$ is

$$\begin{aligned}(S^T S)_{ij} &= \sum_{k=1}^{K} s_i(k\Delta t) s_j(k\Delta t) \\ &= K/2; \, i = j \\ &= 0; \, i \neq j \end{aligned} \tag{5.24}$$

The fundamental frequency components are given by

$$\hat{Y}_c = \frac{2}{K} \sum_{k=1}^{K} y_k \cos(k\theta) \tag{5.25}$$

$$\hat{Y}_s = \frac{2}{K} \sum_{k=1}^{K} y_k \sin(k\theta) \tag{5.26}$$

while for the pth harmonic

$$\hat{Y}_c^{(p)} = \frac{2}{K} \sum_{k=1}^{K} y_k \cos(pk\theta) \tag{5.27}$$

$$\hat{Y}_s^{(p)} = \frac{2}{K} \sum_{k=1}^{K} y_k \sin(pk\theta) \tag{5.28}$$

where $\theta = 2\pi/K$. As in the Kalman filter case in Section 3.9 under the assumption of a constant variance error model and the external removal of the offset, the DFT is an optimal estimate of the fundamental and of all of the harmonics permitted by the sampling rate. The estimates of the harmonics given in Equations (5.27) and (5.28) are not used in line relaying but do play a role in transformer protection, as will be seen in Chapter 6.

The concept of the frequency response of algorithms which take real samples and produce a complex output is subtle. Both Equations (5.25) and (5.26) have frequency responses shown in Figure 5.10a. Both frequency responses have even magnitudes (as a function of ω) and odd phases. The cosine channel is superior for frequencies above the fundamental, while the sine is better below the fundamental. Both parts are required to compute the complex phasor output. A filter with a real input but having a complex output would have a frequency response such as shown

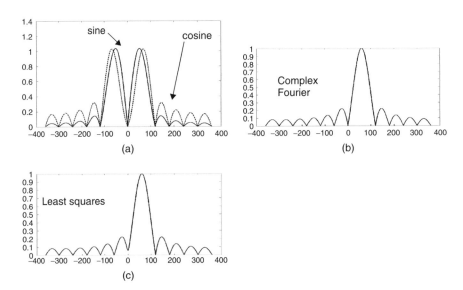

Figure 5.10 Frequency response of full-cycle algorithms at 12 samples per cycle. (a) Sine and Cosine. (b) Complex Fourier. (c) Least squares fitting

in Figure 5.10b. Note the frequency response does not have an even magnitude and odd phase. The frequency response of the least squares algorithm is shown in Figure 5.10c for a sampling rate of 12 samples a cycle. It can be seen that the Fourier algorithm rejects DC while the least squares algorithm does not. The fitting algorithm does reject the exponential it was designed to reject, of course, and has the advantage that no external filtering is required to eliminate the exponential. The disadvantage is that the computation is quite a bit more complicated. We will see in a later section that the Fourier algorithm can be made particularly simple by a suitable choice of the sampling frequency. The additional filtering provided by a longer window is obvious if Figure 5.10 is compared with Figure 5.8.

5.3.3 Fourier algorithms with shorter windows

If only the fundamental frequency components are included in the signal set $\{s_n(t)\}$, the Fourier-like calculations can be carried out for windows of any length. The difference is that the matrix $(S^T S)$ is no longer diagonal. It is only a 2×2 matrix, however. Equation (5.23) becomes:

$$
\begin{bmatrix} \hat{Y}_c \\ \hat{Y}_s \end{bmatrix} = \begin{bmatrix} \sum_{k=1}^{K} \cos^2(k\theta) & \sum_{k=1}^{K} \cos(k\theta)\sin(k\theta) \\ \sum_{k=1}^{K} \cos(k\theta)\sin(k\theta) & \sum_{k=1}^{K} \sin^2(k\theta) \end{bmatrix}^{-1} \begin{bmatrix} \sum_{k=1}^{K} y_k \cos(k\theta) \\ \sum_{k=1}^{K} y_k \sin(k\theta) \end{bmatrix}
$$

$$(5.29)$$

These are the estimates of Section 3.9 obtained from the Kalman filter under the assumptions of no initial estimate and a constant covariance matrix for the measurement error. With $K = 3$ they are a shifted version of Equations (5.4) and (5.5), and with $\theta = 2\pi/K$ they correspond to the full-cycle Fourier. With an even number of samples per half-cycle, Equation (5.29) generates the half-cycle Fourier algorithm.[7]

$$
\hat{Y}_c = \frac{2}{K} \sum_{k=1}^{K} y_k \cos(k\theta) \tag{5.30}
$$

$$
\hat{Y}_s = \frac{2}{K} \sum_{k=1}^{K} y_k \sin(k\theta) \tag{5.31}
$$

The frequency response of the half-cycle Fourier algorithm is shown in Figure 5.11. It can be seen that the shorter window algorithm still rejects the odd harmonics but has lost the ability to reject even harmonics. The DC response is also poor. The algorithm was intended to be used with an external filter to eliminate the exponential offset in the current.

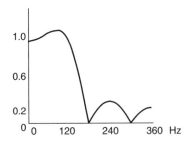

Figure 5.11 Frequency response of half-cycle Fourier algorithm, 12 samples per cycle

5.3.4 *Recursive forms*

The Fourier calculations in Equations (5.25) and (5.26) or Equations (5.30) and (5.31) represent more calculations than are actually necessary in practice. In addition, the computed phasor rotates as in Equation (5.13). If we examine the complex form for the computation involving samples ending at L and drop the factor of 2/K in Equations (5.25) and (5.26) or Equations (5.30) and (5.31),

$$Y^{(L)} = \sum_{k=L-K+1}^{L} y_k e^{-j(k+K-L)\theta}$$

and rotate by an angle of $(K-L)\theta$ to keep the result stationary

$$\tilde{Y}^{(L)} = Y^{(L)} e^{j(K-L)\theta} = \sum_{k=L-K+1}^{L} y_k e^{-jk\theta} \tag{5.32}$$

and

$$\tilde{Y}^{(L-1)} = \sum_{k=L-K}^{L-1} y_k e^{-jk\theta} \tag{5.33}$$

The difference between Equations (5.32) and (5.33) is the last term of (5.32) and the first term of Equation (5.33), i.e.

$$\tilde{Y}^{(L)} = \tilde{Y}^{(L-1)} + [y_L e^{-jL\theta} - y_{L-K} e^{-j(L-K)\theta}]$$

or

$$\tilde{Y}^{(L)} = \tilde{Y}^{(L-1)} + [y_L - y_{L-K} e^{jK\theta}] e^{-jL\theta} \tag{5.34}$$

Equation (5.34) is valid for any length window (any K). If $K\theta$ is not a multiple of a half-cycle then the real and imaginary parts of Equation (5.34) must be multiplied by the matrix $(\mathbf{S^TS})^{-1}$ to form the estimates. For the full-cycle window, $K\theta = 2\pi$ and the recursive form of the full-cycle algorithm becomes

$$\hat{Y}_c^{(new)} = \hat{Y}_c^{(old)} + [y_{new} - y_{old}]\cos(L\theta) \tag{5.35}$$

$$\hat{Y}_s^{(new)} = \hat{Y}_s^{(old)} + [y_{new} - y_{old}]\sin(L\theta) \tag{5.36}$$

where y_{new} is the newest sample corresponding to L and y_{old} is the oldest sample corresponding to a full cycle earlier. If the signal is a purely periodic signal $y_{new} = y_{old}$ and the phasor does not change. Notice also that only one multiplication and one addition are required to accomplish the update of the real and imaginary parts of the phasor estimate. The half-cycle versions of Equations (5.35) and (5.36) with $K\theta = \pi$ are

$$\hat{Y}_c^{(new)} = \hat{Y}_c^{(old)} + [y_{new} + y_{old}]\cos(L\theta) \tag{5.37}$$

$$\hat{Y}_s^{(new)} = \hat{Y}_s^{(old)} + [y_{new} + y_{old}]\sin(L\theta) \tag{5.38}$$

where y_{new} and y_{old} are a half cycle apart in Equations (5.37) and (5.38). Again if $y_{new} = y_{old}$ (a pure fundamental frequency signal or a signal with only odd harmonics) the phasor is unchanged. The missing factor of 2/K may be included in the scaling of the numbers in the microprocessor.

Although the recursive forms of the Fourier calculations are particularly simple, it must be recognized that a large number of currents and voltages must be processed for full three-phase protection of a transmission line (see Section 5.4). The calculations can be further simplified by clever choice of the sampling rate. A sampling rate of four times a cycle would only require the sines and cosines of multiples of 90° (i.e. 0, ±1) and hence make the update a pure addition. Unfortunately this is a rather slow sampling rate for some applications. If the rate is increased to eight times a cycle, sines and cosines of multiples of 45° are required (0, ±1 and $\pm\sqrt{2}/2$). At 12 times a cycle, the multipliers are 0, ±1, ±1/2, and $\sqrt{3}/2$. In each of the latter cases, the multiplication by an irrational number can be accomplished externally by an analog voltage divider, or approximated in software by a number of shifts and adds.

Example 5.2

An approximation to $\sqrt{3}/2$ can be obtained for a 16 bit processor as

$$\sqrt{3}/2 = 2^{-1} + 3 \times 2^{-3} - 2^{-7} - 2^{-10} - 3 \times 2^{-14}$$

$$0.8660254 \cong 0.8660278$$

The error of 0.0000024 is less than 2^{-16}. In fact, it is not clear that the first three or four terms are not adequate. The implementation of the five term approximation then involves a sequence of shifts and adds and, in one implementation, is done while the next sample is being converted by the A/D converter.

5.3.5 Walsh function algorithms

The Walsh functions have the advantage that all of the multiplications required in the evaluation of Equation (5.23) are ± 1. If we take $N = 2^n$ and let

$$s_n(t) = w_n(t) \tag{5.39}$$

where the Walsh functions are defined as in Section 3.3, then

$$(S^T S)^{-1} = 2^{-n} I \tag{5.40}$$

$$\hat{Y}_n = \frac{1}{2^n} \sum_{k=1}^{2^n} y_k w_n(k\theta) \tag{5.41}$$

Example 5.3

The first four Walsh functions are shown in Figure 5.12 along with the sample values $w_n(k\theta)$. An expansion in terms of the first four Walsh functions would involve the calculations

$$\hat{Y}_1 = \frac{1}{4}(y_1 + y_2 + y_3 + y_4)$$

$$\hat{Y}_2 = \frac{1}{4}(y_1 + y_2 - y_3 - y_4)$$

$$\hat{Y}_3 = \frac{1}{4}(y_1 - y_2 - y_3 + y_4)$$

$$\hat{Y}_4 = \frac{1}{4}(y_1 - y_2 + y_3 - y_4)$$

The only problem with the Walsh expansion is that explained in Section 3.3. A large number of Walsh terms must be included in order to obtain an accurate estimate of the components. The advantage of the simplicity of Equation (5.41) is counterbalanced by the need for a large number of terms and the need to convert from Walsh to Fourier. The frequency response of the Walsh algorithm

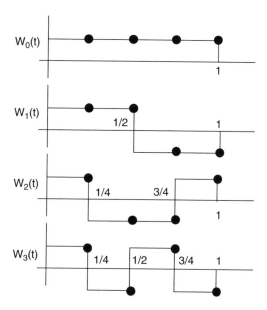

Figure 5.12   The first four Walsh functions and their sample values

is indistinguishable from the full-cycle Fourier algorithm if a sufficient number of Walsh coefficients are used.

5.3.6 Differential-equation algorithms

The differential-equation algorithms represent a second major theme in line relaying. The algorithms described so far in this chapter are based on a description of the waveforms and are essentially impedance relay algorithms. These algorithms attempt to estimate the fundamental frequency components of currents and voltages in order to compute the impedance to the fault. The differential-equation algorithms, on the other hand, are based on a model of the system rather than on a model of the signal. They can still be regarded as parameter estimation, however. If we take the single-phase model of the faulted line shown in Figure 5.2 and write the differential-equation relating the voltage and current seen by the relay, we obtain

$$v(t) = R\,i(t) + L\frac{di(t)}{dt} \tag{5.42}$$

Since both $v(t)$ and $i(t)$ are measured, it seems possible that we can estimate the parameters R and L and hence the distance to the fault. Since derivatives of measured quantities are difficult to produce, McInnes and Morrison obtained a more tractable form of Equation (5.42) by integrating the Equation (5.42) over two consecutive intervals:[8]

$$\int_{t_0}^{t_1} v(t)dt = R \int_{t_0}^{t_1} i(t)dt + L[i(t_1) - i(t_0)]$$ (5.43)

$$\int_{t_1}^{t_2} v(t)dt = R \int_{t_1}^{t_2} i(t)dt + L[i(t_2) - i(t_1)]$$ (5.44)

The integrals in Equations (5.43) and (5.44) must be approximated from the sample values. If the samples are equally spaced at an interval Δt and the trapezoidal rule is used for the integrals, viz.

$$\int_{t_0}^{t_1} v(t)dt = \frac{\Delta t}{2}[v(t_1) + v(t_0)] = \frac{\Delta t}{2}[v_1 + v_2]$$

then Equations (5.43) and (5.44) can be written for samples at k, k+1, and k+2 as

$$\begin{bmatrix} \frac{\Delta t}{2}(i_{k+1} + i_k) & (i_{k+1} - i_k) \\ \frac{\Delta t}{2}(i_{k+2} + i_{k+1}) & (i_{k+2} - i_{k+1}) \end{bmatrix} \begin{bmatrix} R \\ L \end{bmatrix} = \begin{bmatrix} \frac{\Delta t}{2}(v_{k+1} + v_k) \\ \frac{\Delta t}{2}(v_{k+2} + v_{k+1}) \end{bmatrix}$$

The three samples of current and voltage are sufficient to compute estimates of R and L as

$$R = \left[\frac{(v_{k+1} + v_k)(i_{k+2} - i_{k+1}) - (v_{k+2} + v_{k+1})(i_{k+1} - i_k)}{(i_{k+1} + i_k)(i_{k+2} - i_{k+1}) - (i_{k+2} + i_{k+1})(i_{k+1} - i_k)} \right]$$ (5.45)

$$L = \frac{\Delta t}{2} \left[\frac{(i_{k+1} + i_k)(v_{k+2} + v_{k+1}) - (i_{k+2} + i_{k+1})(v_{k+1} + v_k)}{(i_{k+1} + i_k)(i_{k+2} - i_{k+1}) - (i_{k+2} + i_{k+1})(i_{k+1} - i_k)} \right]$$ (5.46)

Equations (5.45) and (5.46) can be roughly compared with Equations (5.4) and (5.5), the three sample phasor calculations. Both the current and voltage phasors must be computed using Equations (5.4) and (5.5) and then a complex division performed to obtain the impedance to the fault. Equations (5.45) and (5.46) represent the total computation involved in processing the six samples (three current and three voltage samples). In total, the six multiplications and two real divisions in Equations (5.45) and (5.46) compare favorably with computation of Equations (5.4) and (5.5) for current and voltage followed by a complex division which would also involve six multiplications and two real divisions in the form

$$\frac{a + jb}{c + jd} = \frac{(ac + bd)}{c^2 + d^2} + j\frac{(bc - ad)}{c^2 + d^2}$$

The actual difference in speed would depend on the angle θ in Equations (5.4) and (5.5).

The algorithm given by Equations (5.45) and (5.46) is the differential-equation counterpart of the short window algorithm given by Equations (5.6) and (5.7) and as such is not as selective as longer window algorithms. The extension of the differential-equation approach to a longer window takes a number of forms. One approach is to make the intervals $[t_o, t_1]$ and $[t_1, t_2]$ in Equations (5.43) and (5.44) longer. It is possible to select the intervals so that certain harmonics are rejected.[9] If the intervals contain a number of samples, the trapezoidal approximation to each of the integrals will contain sums of a number of samples. Another approach is to do the trapezoidal integration over the interval between adjacent samples and obtain an over defined set of equations in the form

$$\begin{bmatrix} \frac{\Delta t}{2}(i_{k+1} + i_k) & (i_{k+1} - i_k) \\ \frac{\Delta t}{2}(i_{k+2} + i_{k+1}) & (i_{k+2} - i_{k+1}) \\ \vdots & \vdots \\ \frac{\Delta t}{2}(i_{k+N} + i_{k+N-1}) & (i_{k+N} - i_{k+N-1}) \end{bmatrix} \begin{bmatrix} R \\ L \end{bmatrix} = \begin{bmatrix} \frac{\Delta t}{2}(v_{k+1} + v_k) \\ \frac{\Delta t}{2}(v_{k+2} + v_{k+1}) \\ \vdots \\ \frac{\Delta t}{2}(v_{k+N} + v_{k+N-1}) \end{bmatrix}$$

The least squares solution of the over defined equations would involve a large number of multiplications in forming the equivalent of the matrix $(\mathbf{S^T S})$. To avoid some of these problems, a technique of using a sequence of estimates, each of which is obtained from the three-sample algorithm, has been developed.[10] Suppose a region of the R-L plane is chosen to correspond to zone-1 protection of the line. If the result of the three-point calculation Equations (5.45) and (5.46) lies in the characteristic, a counter is indexed by one. If the computed values lie outside the characteristic, the counter is reduced by one. With a threshold of four on the counter, i.e. a trip signal cannot be issued unless the counter is at four, a minimum of six consecutive samples is required. The window length is thus increased by increasing the counter threshold.

Although ingenious, the counting schemes are difficult to compare with other long window algorithms in terms of frequency response. An additional problem in giving the frequency response of the differential-equation algorithms is that there are two signals involved, i(t) and v(t). To obtain some comparison, Figure 5.13 is a frequency response obtained for the average of consecutive three-sample results

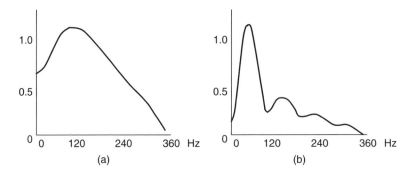

Figure 5.13 Frequency response of the average of consecutive three-sample calculations for Equations (5.45) and (5.46). (a) One-half cycle window at 12 samples per cycle. (b) Full-cycle window at 12 samples per cycle

of Equations (5.45) and (5.46) spanning half-cycle and full-cycle windows at a sampling rate of 12 samples per cycle.

In computing Figure 5.13, it was assumed that the current was the true fundamental frequency current but that the voltage signal varied in frequency. In essence, Figure 5.13 is the frequency response of the averaged numerator of Equations (5.45) and (5.46) with the denominator held at the fundamental frequency values. The frequency response in Figure 5.13 can only roughly be compared with the earlier responses. A different response is produced, for example, if the averaging is done only for non-overlapping three-sample results. An additional problem with the concept of frequency response for these algorithms is that as long as v(t) and i(t) satisfy the differential-equation (no matter what frequencies are present in v(t) and i(t)) then the computed values of R and L are correct.

5.3.6.1   Error analysis for differential-equation algorithms

The property identified in describing the frequency response of the differential-equation algorithms, i.e. that the correct R and L are obtained as long as v(t) and i(t) satisfy the differential equation, is a major strength of these algorithms. The exponential offset in the current satisfies the equation if the correct values of R and L are used, so that it need not be removed. We have seen in Section 5.2 that the voltage seen by the relay just after fault inception has non-fundamental frequency components caused by the power system itself. If the faulted line can accurately be modeled as a series R-L line, then the current will respond to these non-fundamental frequency components in the voltage (according to the differential equation) and no errors in estimating R and L will result. As long as the R-L representation is correct, then the only sources of error are in the measurement of v(t) and i(t). Transducer errors, A/D errors, and errors in the anti-aliasing filters contribute portions of the measured voltage and current which do not satisfy the differential equation and which will cause errors in the estimates. The frequency

response in Figure 5.13 represents the response of the algorithm to such error signals
in the measured voltage. To examine the effect of such errors in both voltage and
current, let the measured current and voltage be denoted by $i_m(t)$ and $v_m(t)$ where

$$i_m(t) = i(t) + \varepsilon_i(t) \tag{5.47}$$

$$v_m(t) = v(t) + \varepsilon_v(t) \tag{5.48}$$

and $v(t)$ and $i(t)$ satisfy the differential equation. The current $i_m(t)$ satisfies the
differential equation

$$Ri_m(t) + L\frac{di_m(t)}{dt} = v(t) + R\varepsilon_i(t) + L\frac{d\varepsilon_i(t)}{dt} \tag{5.49}$$

and the current $i_m(t)$ and voltage $v_m(t)$ are related by:

$$i_m(t) + L\frac{di_m(t)}{dt} = v_m(t) + R\varepsilon_i(t) + L\frac{d\varepsilon_i(t)}{dt} - \varepsilon_v(t) \tag{5.50}$$

The measured voltage and current then satisfy a differential equation with an error
term which is made up of the voltage error plus a processed current error term. The
latter is similar to the error obtained at the output of a mimic circuit with the current
error as an input (see Figure 5.16). Figure 5.13 then can finally be interpreted as
the response of the algorithm to the entire error term in Equation (5.50), assuming
that the sum of the last three terms on the right hand side of the equation are
thought of as a signal $\cos(\omega t)$; the last three terms in (5.50) also make it clear that
error signals that satisfy the differential equation do not contribute a net error to
Equation (5.50).

There is an additional subtlety in the three-sample differential-equation algorithm.
The denominator of Equations (5.45) and (5.46) is not a constant but rather a function of time which has maxima and minima. The denominator can be simplified to

$$(i_{k+1} + i_k)(i_{k+2} - i_{k+1}) - (i_{k+2} + i_{k+1})(i_{k+1} - i_k) = -2(i_{k+1}^2 - i_k i_{k+2})$$

If we assume that

$$i_{k+1} = I\cos(\omega_o t) - I\cos(\omega_o t_o)\,e{-}\frac{R}{L}(t - t_o)$$

and use 12 samples per cycle ($\theta = 30°$) with a line time constant of 40 ms, then

$$2\left(i_{k+1}^2 - i_k i_{k+2}\right) = I^2\left[0.5 - 0.5384\cos(\omega_o t + 7.41°)\cos(\omega_o t_o)\,e^{-\frac{R}{L}(t - t_o)}\right]$$

The denominator is shown in Figure 5.14 as a function of the time of the kth
sample for the case of maximum offset ($\omega_0 t_0 = 0$). When the denominator is small

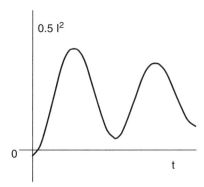

Figure 5.14 The denominator of Equations (5.45) and (5.46) for maximum offset in the current

the error terms from Equation (5.50) are amplified. In the limit as the denominator becomes zero the estimates are unacceptably sensitive to even the smallest error terms. A counting algorithm will deal with such poor estimates by indexing down because the estimate is not in the characteristic. The net effect is then only a delay in issuing the trip signal. Since the denominator is a constant if the offset is absent, it can be seen that the supposed immunity of the differential-equation algorithms to offset is a bit of myth.

The preceding has assumed that the actual voltage and current satisfy the differential equation but that there are errors made in the measurement process. As seen in Section 5.2, the largest contributions to errors in the waveform algorithms are non-fundamental frequency signals from the power system itself. If the differential-equation algorithms were immune to these signals, there would be much to recommend them. A somewhat more realistic model of the faulted line can be used to investigate the impact of these power system signals on such algorithms. The circuit shown in Figure 5.15 includes the shunt capacitance of the transmission line at the relay terminals. The voltage source v(t) represents the voltages seen in Figure 5.5 made up of a fundamental plus non-fundamental components whose frequency and phase are unpredictable. The current i(t) is the current measured by the relay. The

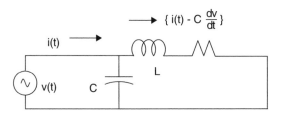

Figure 5.15 A single-phase line model with shunt capacitance

actual relationship between the measured voltage and current is given by

$$v(t) = Ri(t) + L\frac{di(t)}{dt} - RC\frac{dv(t)}{dt} - LC\frac{d^2v(t)}{d^2t} \tag{5.51}$$

If the algorithm of Equations (5.45) and (5.46) is used then the last two terms in Equation (5.51) must be regarded as error terms. The magnitude of these terms is a function of fault location and of the frequency of the signals in v(t). The dependence on fault location is quadratic, reaching a maximum for a fault at the end of the line. For faults at the far end of a long high-voltage line, these terms can be quite substantial, especially if high frequencies are included in v(t). On the other hand, the terms RC and LC are small for close-in faults and on lower voltage lines.

A solution to this problem, of course, is to include the capacitance in the system model. This has been proposed and central differences have been used to approximate the derivatives in Equation (5.51).[11] It seems desirable to integrate Equation (5.51) once to recover some similarity with Equations (5.45) and (5.46). If this is done and

$$\left[\frac{dv(t)}{dt}\right]_{t_k} \cong \frac{v_k - v_{k-1}}{\Delta t}$$

then a more elaborate version of Equations (5.45) and (5.46) can be obtained by considering four consecutive intervals

$$\begin{bmatrix} \frac{\Delta t}{2}(i_{k+1} + i_k) & (i_{k+1} - i_k) & -(v_{k+1} - v_k) & \frac{-1}{\Delta t}(v_{k+1} - 2v_k + v_{k-1}) \\ \frac{\Delta t}{2}(i_{k+2} + i_{k+1}) & (i_{k+2} - i_{k+1}) & -(v_{k+2} - v_{k+1}) & \frac{-1}{\Delta t}(v_{k+2} - 2v_{k+1} + v_k) \\ \frac{\Delta t}{2}(i_{k+3} + i_{k+2}) & (i_{k+3} - i_{k+2}) & -(v_{k+3} - v_{k+2}) & \frac{-1}{\Delta t}(v_{k+3} - 2v_{k+2} + v_{k+1}) \\ \frac{\Delta t}{2}(i_{k+4} + i_{k+3}) & (i_{k+4} - i_{k+3}) & -(v_{k+4} - v_{k+3}) & \frac{-1}{\Delta t}(v_{k+4} - 2v_{k+3} + v_{k+2}) \end{bmatrix}$$

$$\times \begin{bmatrix} R \\ L \\ RC \\ LC \end{bmatrix} = \begin{bmatrix} \frac{\Delta t}{2}(v_{k+1} + v_k) \\ \frac{\Delta t}{2}(v_{k+2} + v_{k+1}) \\ \frac{\Delta t}{2}(v_{k+3} + v_{k+2}) \\ \frac{\Delta t}{2}(v_{k+4} + v_{k+3}) \end{bmatrix} \tag{5.52}$$

If Equation (5.52) is thought of as a partitioned matrix in the form

$$\begin{bmatrix} M_{11} & M_{12} \\ M_{21} & M_{22} \end{bmatrix}\begin{bmatrix} p \\ Cp \end{bmatrix} = \begin{bmatrix} r_1 \\ r_2 \end{bmatrix} \tag{5.53}$$

where $\mathbf{p}$ is the parameter vector made up of R and L, then the estimate can be formed by solving the second set of equations and substituting into the first set to obtain:[11]

$$\begin{bmatrix} \hat{R} \\ \hat{L} \end{bmatrix} = \left(M_{11} - M_{12}\, M_{22}^{-1}\, M_{21} \right)^{-1} \left(r_1 - M_{12}\, M_{22}^{-1}\, r_2 \right) \qquad (5.54)$$

Equation (5.54) represents a formidable amount of computation given that the matrices involved are formed of measured voltages and currents. It is not clear that the computation is warranted since the pi-section representation is still an approximation to the transmission line model.

5.3.7 Kalman filter algorithms

In 1981, simulation experiments performed on a 345 kV line connecting a generator and a load led to the conclusion that the covariance of the noise in the voltage and current was not a constant but, rather, decayed in time. If the time constant of the decay is comparable to the decision time of the relay then the Kalman filter solution of Section 3.9 is appropriate for the estimation problem. The voltage was modeled as in Equations (3.83) and (3.84) with

$$X = \begin{bmatrix} Y_c \\ Y_s \end{bmatrix}, \varphi = \begin{bmatrix} 1 & 0 \\ 0 & 1 \end{bmatrix}, \text{ and } H_k = [\cos(k\theta)\ \sin(k\theta)] \qquad (5.55)$$

while a three-state model was used for the current to account for the offset. In particular, the current was assumed to be described by

$$x = \begin{bmatrix} Y_c \\ Y_s \\ Y_0 \end{bmatrix}, \varphi = \begin{bmatrix} 1 & 0 & 0 \\ 0 & 1 & 0 \\ 0 & 0 & e^{-\beta\Delta t} \end{bmatrix} \qquad (5.56)$$

and

$$H_k = [\cos(k\theta)\ \sin(k\theta)\, 1] \qquad (5.57)$$

In both cases the covariance of the measurement noise was taken to be

$$R_k = K\, e^{-k\Delta t/T} \qquad (5.58)$$

where K was different for voltage and current models and T was one-half the expected time constant of the protected line. Using the error model given by Equation (5.58) and a prior estimate for the state, the Kalman filter is required as shown in Section 3.9.

The Kalman-type description has been extended to an 11 state model including two states for each of the fundamental through the fifth harmonic plus one state for the offset.[15] A constant error covariance was assumed, however, so that the only justification for the Kalman filter as opposed to least squares or Fourier techniques is the prior estimate. It should be recognized that either Kalman filter approach is a waveform algorithm, in that the fundamental frequency phasors are being estimated. Exact prior estimates of the fundamental frequency phasors are equivalent to prior knowledge of the fault location and are not reasonable. Crude prior estimates are obviously available (for example, the fault is within 1000 miles of the bus) but such crude estimates will be quickly overcome by the measurements. The Kalman filter approach cannot be justified simply to include a questionable prior estimate of the fault location. It must be accepted that the only reason to use the Kalman filter approach is to account for the time-varying measurement error model given by Equation (5.58).

In examining the simulation model used to obtain Equation (5.58),[12-14] the absence of any network structure at the source bus is an obvious limitation. The non-fundamental frequency terms described in Section 5.2 are produced by the network structure. A line connected directly to a high short circuit capacity source will not see these non-fundamental frequency terms. Other more detailed studies[15] and field data[16] support the conclusion that, although the error term decays, it does so at a much slower rate than suggested by Equation (5.58) (for example, see Figure 5.5). In summary, the Kalman filter approach is an alternate technique for estimating the fundamental frequency components of voltage and current involving considerably more computation than the alternatives, which is justified only if the measurement error covariance decays significantly during the first cycle after the fault.

5.3.8 Removal of the DC offset

One of the distinctions between the preceding algorithms is the treatment of the offset in the current. The short window Fourier algorithms in particular require that the offset be removed prior to processing, while the differential-equation algorithms ideally do not require its elimination. Consider the R-L line model of Figure 5.16. If the fault occurs at $t = 0$, while the voltage of the source is

$$e(t) = \sqrt{2\,E}\cos(\omega_0 t + \phi) \tag{5.59}$$

the fault current $i(t)$ (assuming no pre-fault current in the circuit) is given by

$$i(t) = \frac{\sqrt{2\,E}}{\sqrt{(R^2 + X^2)}}\cos(\omega_0 t + \phi - \Psi) - \frac{\sqrt{2\,E}}{\sqrt{(R^2 + X^2)}}\cos(\phi - \Psi)\,e^{\frac{\omega_0 R t}{X}} \tag{5.60}$$

The second term in the expression for $i(t)$ is the decaying DC component. The traditional method of eliminating the DC component is with the use of a mimic

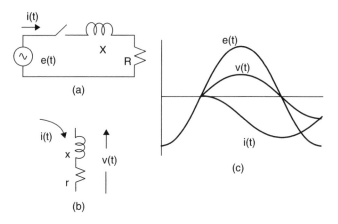

Figure 5.16 Generation of exponentially decaying DC offset and its removal. (a) Fault circuit. (b) Mimic circuit. (c) Voltage and current waveforms

circuit as shown in Figure 5.16 (b). The fault current I (or its analog in the secondary winding of a current transformer) is passed through a burden of $(r + jx)$ where $x/r = X/R$. In this case, the voltage across the burden $(r + jx)$ is given by

$$v(t) \frac{\sqrt{2}\,E}{\sqrt{(R^2 + X^2)}} \sqrt{(r^2 + x^2)} \cos(\omega_0 t + \Psi) \qquad (5.61)$$

This voltage $v(t)$ is proportional to the fault current and is without the DC offset. It can be used to represent the fault current. The phase angle of $v(t)$ is different from that of $i(t)$ Indeed, $v(t)$ is in phase with $e(t)$, the voltage that produced the fault current.

Figure 5.16 (c) shows the waveforms of the currents and voltages associated with the fault circuit and the mimic circuit. The burden $(r + jx)$ is known as the mimic impedance, since it mimics the fault path impedance. Such circuits are commonly used in analog and digital relays.[7]

It should be noted that the r and x of the mimic impedance must be set to specific values, whereas the fault circuit X/R may change depending upon network switching or fault arc resistance. In addition, any noise (or extraneous high frequency components) present in the current tends to be amplified by the mimic circuit. In the case of computer relays, the low-pass anti-aliasing filter attenuates a significant part of the noise.

As microcomputer capabilities have improved, it has become possible to include the mimic circuit representation within the microcomputer.[17] Consider the samples of $i(t)$ taken at instants $k\Delta t$, $\{k = 1, 2, 3 \ldots\}$. We may write

$$i_k = Y_o \gamma^k + Y_s \sin k\theta + Y_c \cos k\theta \qquad (5.62)$$

where, from Equation (5.60),

$$Y_o = -\frac{\sqrt{2}\,E}{\sqrt{R^2 + X^2}}\cos(\phi - \Psi)$$

$$Y_s = -\frac{\sqrt{2}\,E}{\sqrt{R^2 + X^2}}\sin(\phi - \Psi) \qquad (5.63)$$

$$Y_c = -\frac{\sqrt{2}\,E}{\sqrt{R^2 + X^2}}\cos(\phi - \Psi)$$

$$\gamma = e^{\frac{-\theta R}{X}}$$

Assuming that the fault circuit X/R is known, γ in Equation (5.63) is known; and Equation (5.62) for n samples (n>3) can be written as an over-determined set of equations:

$$i = \begin{bmatrix} i_1 \\ i_2 \\ \vdots \end{bmatrix} = \begin{bmatrix} \gamma \sin\theta \sin 2\theta .. \\ \gamma^2 \cos\theta \cos 2\theta .. \end{bmatrix} \begin{bmatrix} Y_o \\ Y_s \\ Y_c \end{bmatrix} \qquad (5.64)$$

or,

$$\mathbf{i = JY}$$

from which the unknown vector **Y** can be determined:

$$Y = (J^T J)^{-1} J^T i \qquad (5.65)$$

The constants Y_0, Y_s, Y_c having been found, the samples of v(t) can also be found. The k'th sample of v(t) is given by

$$v_k = \frac{\sqrt{2}\,E}{\sqrt{(R^2 + X^2)}}\sqrt{(r^2 + x^2)}\cos(k\theta + \varphi) \qquad (5.66)$$

We may combine Equations (5.65) and (5.66) into a single algorithm for finding v_k from i_k:

$$v = G(J^T J)^{-1} J^T [i - Y_o(\gamma, \gamma^2, \ldots)^T] \qquad (5.67)$$

where

$$G = \sqrt{(r^2 + x^2)} \begin{bmatrix} \sin(\theta + \psi)\sin(2\theta + \psi).. \\ \cos(\theta + \psi)\cos(2\theta + \psi).. \\ \cdot \quad\quad \cdot \quad\quad .. \\ \cdot \quad\quad \cdot \quad\quad .. \end{bmatrix} \qquad (5.68)$$

With three data points, Equation (5.67) has no redundancy. With a greater number of data points, the redundancy becomes favorable and the mimic circuit representation becomes less immune to noise. The reference cited earlier gives examples of laboratory experiments with three-point and six-point algorithms.[17] It has been shown that real-time elimination of DC offset by digital techniques is feasible, and produces results comparable to those obtained with analog mimic circuits.

5.4 Beyond parameter estimation

There is a great deal more involved in a line relay than the estimation of the parameters. It has been seen in Section 5.3 that all of the line relaying algorithms can be thought of as attempts to estimate the parameters of the signal or of the model of the faulted line. Comparisons of relaying algorithms based solely on their ability to estimate these parameters are fraught with difficulty. Different assumptions about the measurement errors produce different optimal estimates. Different network models used in computer simulations produce different non-fundamental frequency terms in the post-fault waveforms. Any algorithm can be made to circumvent a particular deficiency by putting auxiliary features in the total relaying program. For example, the short window Fourier algorithms are made insensitive to the offset through the use of an external mimic impedance or its equivalent algorithm. Comparisons of algorithms which do not recognize these auxiliary features can lead to absurd results.

A more complete comparison requires an understanding of the total relaying program. Some of the issues which must be considered beyond the parameter estimation are: determination of the fault type and how the fault type impacts the parameter estimation; monitoring of the quality of the estimates; and speed-reach considerations. Each of these issues will be examined in this section in a general setting. Finally, one particular algorithm will be presented in more detail as an example.

5.4.1 Relay programs based upon fault classification

The algorithms presented in the preceding sections are essentially single-phase algorithms. The waveform algorithms are for a single current or voltage and the differential-equation algorithms are for a single-phase R-L model. A line relay using one of these algorithms would protect one terminal of a three-phase transmission line. If the phases are labeled as a, b, c then there are a total of ten possible faults that can be seen by the relay. They are

  a – ground
  b – ground
  c – ground

$$a - b$$
$$a - c$$
$$b - c$$
$$a - b - ground$$
$$a - c - ground$$
$$b - c - ground$$
$$a - b - c$$

Since the fault type is not known *a-priori* the algorithm must process samples of all the voltages and currents in order to determine the fault type.

Conceptually, the simplest distance relaying algorithm would process six single-phase distance Equations corresponding to three phase-ground faults and three phase-phase faults. Early computers – and even modern microprocessors – would be hard pressed to process all six distance Equations between successive samples. It may be that the next generation of microprocessors will make such a procedure practical. The overall flow chart of a relay program of this type is shown in Figure 5.17. When a sample set of voltages and currents are acquired, the fault distance is estimated by assuming each of the six possible types (see Section 2.3). The distance may be calculated by phasors or through the corresponding single-phase differential equation. Once the six distances are computed, each is checked against the relay characteristic to determine the appropriate relay response. It should be noted that, for a given fault, only some of the distance calculations will produce correct fault distances. Thus for a phase *a-b* fault, block *1* in Figure 5.17 will produce the correct fault distance, while all other equations will produce distances which lie outside the relay zones. Similarly, a phase *a-b-g* fault will produce correct fault distances in blocks *1, 4* and *5*. A three-phase fault will produce correct distances in all six blocks.

A considerable saving in computation can be achieved by utilizing a first step which attempts to determine the fault type. This principle is illustrated in Figure 5.18. In general, only one of the six equations would be processed for any fault, and considerable computational efficiency would be achieved.

An early classification program used a voltage deviation criterion to classify a fault.[18] Thus, for a phase *a-g* fault, the voltage of phase *a* should deviate from its value one cycle ago. Such a comparison is made, and if it is found to persist (while other phase voltages remain undisturbed), an *a-g* fault is declared (see Figure 5.19). In order to guard against declaring a fault when a data sample is corrupted by noise, an accumulated number of points at which the voltage difference exceeds a tolerance may be compared against a threshold. A count is maintained for each phase, and if counts of two phases reach the threshold simultaneously, a phase-to-phase fault is declared. One could supplement a voltage classifier with a similarly designed current classifier.

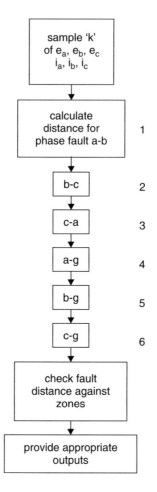

Figure 5.17   A simple relay program

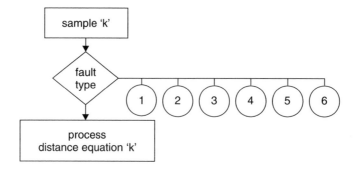

Figure 5.18   A fault classification relay program

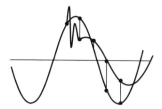

Figure 5.19   Voltage waveforms for a voltage classification algorithm

Fault classification schemes based upon voltages and/or currents are used in several practical relaying programs. They produce correct fault classification under most reasonable conditions but do mis-classify in a significant number of cases. The voltage based classifiers tend to be error-prone when short lines are supplied from a weak system or when a long line is fed from a strong system. In the former case all voltages deviate significantly from nominal while in the latter case faults near a zone boundary produce hardly any voltage deviation in the faulted phase at the relay location. Current based classifiers are similarly confused when load currents are significant compared to fault current. Whenever the classification is uncertain, it must be recognized as such, and all six equations must be processed at each sample time.

Another drawback of the programs using fault classification is that no fault processing can be begun until the classification phase is complete. If voltage deviation counters are to be checked against thresholds, valuable time is lost while the distance calculation (and the relay decision) is held in abeyance waiting for the classification step to be completed. This makes for slower relay response time.

For the differential-equation algorithms, the Clarke[19] components offer another possible approach.[10] The Clarke components with phase a as reference are obtained by multiplying the phase quantities by the matrix

$$\mathbf{T_c} = \frac{1}{3}\begin{bmatrix} 1 & 1 & 1 \\ 2 & -1 & -1 \\ 0 & \sqrt{3} & -\sqrt{3} \end{bmatrix} \tag{5.69}$$

It can be verified by multiplication that $\mathbf{T_c^T}\,\mathbf{T_c}$ is diagonal but is not normalized so that it is not a unitary matrix. The components are referred to as zero, alpha, and beta components. For example, for

a phase-to-ground fault	$I_\alpha = 2\,I_0$ and $I_\beta = 0$
b-c to ground fault	$I_\alpha = -I_0$
b-c fault	$I_\alpha = 0$ and $I_0 = 0$
3-phase fault	$I_0 = 0$

If the Clarke transformation is also performed with b and c phases as reference and if the neutral current I_n is measured, these conditions can be translated into two types depending on whether I_n is zero or not.

If $I_n \neq 0$ (ground faults)

$$I_b - I_c = 0 \qquad\qquad a - \text{ground fault}$$
$$I_a - I_c = 0 \qquad\qquad b - \text{ground fault}$$
$$I_b - I_a = 0 \qquad\qquad c - \text{ground fault}$$
$$2\,I_a - I_b - I_c + I_n = 0 \quad\ \ \text{b-c} - \text{ground fault}$$
$$2\,I_b - I_a - I_c + I_n = 0 \quad\ \ \text{a-c} - \text{ground fault}$$
$$2\,I_c - I_a - I_b + I_n = 0 \quad\ \ \text{a-b} - \text{ground fault}$$

If $I_n = 0$ (phase faults)

$$2\,I_a - I_b - I_c = 0 \quad \text{b-c fault}$$
$$2\,I_b - I_a - I_c = 0 \quad \text{a-c fault}$$
$$2\,I_c - I_a - I_b = 0 \quad \text{a-b fault}$$

If none of the equalities is satisfied for a phase fault, it is assumed that the fault is a three-phase fault. The nine quantities are actually checked against a small threshold rather than zero. It can be seen that the fault must be correctly classified in order to use the correct current and voltage in the algorithm for computing R and L. An obvious problem is mis-classification or an evolving fault. If some samples have been processed under the assumption that the fault is a – ground, for example, and it is recognized that the fault is, in fact, a-b – ground then the counter must be reset and the processing started over. The net effect is a delay in clearing the fault.

5.5 Symmetrical component distance relay

The use of phasor calculations permits the use of symmetrical components in the detection of fault type. The Symmetrical Component Distance Relay[20] overcomes the uncertainty of the fault classifiers and their attendant delay of response in a neat manner; the symmetrical component transformation[20] applies to phasor quantities and with phase a as reference takes the form

$$\mathbf{T}_s = \frac{1}{3} \begin{bmatrix} 1 & 1 & 1 \\ 1 & \alpha & \alpha^2 \\ 1 & \alpha^2 & \alpha \end{bmatrix} \tag{5.70}$$

where α is $e^{j2\pi/3}$.

The resulting component quantities are referred to as zero, positive, and negative quantities, denoted by subscripts 0, 1, and 2 respectively. If we consider a fault at a fraction k of the line length away from the relay with pre-fault load currents

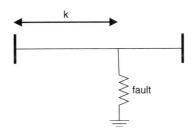

Figure 5.20   A fault at a fraction k of the line length

of I_0, I_1, and I_2 as shown in Figure 5.20, then we can define the changes in the component currents as

$$\Delta I_0 = I_0 - I_0$$

$$\Delta I_1 = I_1 - I_1 \qquad (5.71)$$

$$\Delta I_2 = I_2 - I_2$$

In fact, only I_1 will be significant in practice. The voltage drops in the line can be defined as

$$\Delta E_0 = \Delta I_0 \, Z_0$$

$$\Delta E_1 = \Delta I_1 \, Z_1 \qquad (5.72)$$

$$\Delta E_2 = \Delta I_2 \, Z_2$$

where Z_0, Z_1, and Z_2 are the sequence impedances of the entire line. The ratios

$$k_0 = \frac{E_0}{\Delta E_0}$$

$$k_1 = \frac{E_1}{\Delta E_1}$$

$$(5.73)$$

$$k_2 = \frac{E_2}{\Delta E_2}$$

$$k_l = \frac{Z_1 \bar{I}_1}{\Delta E_1}$$

play an important role in determining fault location for all fault types. For example, a three-phase fault with fault resistance, R_{1f}, would involve only the positive sequence network as shown in Figure 5.21. The voltage equation is:

$$0 = \Delta E_1 [k_l - k(1 + k_l)] - R_{1f} I_{1f}$$

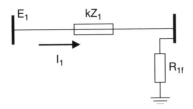

Figure 5.21 Positive sequence network for a three-phase fault

or

$$k = \frac{k_1}{1 + k_1} + \varepsilon_r \tag{5.74}$$

where

$$\varepsilon_r = \frac{-R_{1f}I_{1f}}{\Delta E_1 (1 + k_1)}$$

By considering all possible fault types a general expression for the fractional distance to the fault can be obtained in the form

$$k = \frac{k_1 + k_2 k_2' + k_0 k_0'}{1 + k_0' + k_2' + k_1} \tag{5.75}$$

where k_0' and k_2' play something of the role of the nine conditions developed in the Clarke component case.[20] They are

$$k_0' = \left| \frac{\Delta E_0}{\Delta E_1} \right|$$

and

$$k_2' = \begin{cases} 1 & \text{if } |\Delta E_2| \cong |\Delta E_1| \\ 0 & \text{otherwise} \end{cases}$$

A flow chart for the k calculation is shown in Figure 5.22. The obvious advantage of the k calculation is that it is not necessary to determine the fault type before meaningful calculations can be performed.

5.5.1 SCDFT

The computational burden associated with the computation of the symmetrical components can be greatly reduced by the choice of sampling frequency. If sampling frequencies that correspond to $\theta = 30°, 60°$, or $120°$ are used then the discrete Fourier transform calculation and the symmetrical component calculation can be

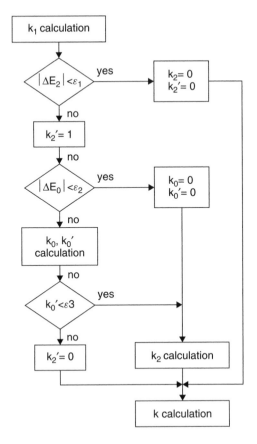

Figure 5.22   The flow chart for the k algorithm

effectively combined. Using $\theta = 30°$, the factor α in Equation (5.70) is seen to be a rotation by $4 \times \theta$ and α^2 a rotation by $8 \times \theta$ or $-4 \times \theta$. If we let

$$\Delta \, y_L = (y_{new} - y_{old}) \text{ for the full-cycle algorithm}$$

and

$$\Delta \, y_L = (y_{new} + y_{old}) \text{ for the half-cycle algorithm}$$

given in Equations (5.35) – (5.38), we can produce the full- or half-cycle versions of the SCDFT (Symmetrical Component Discrete Fourier transform) with a single expression.[7]

$$Y_{OC}^{(L+1)} = Y_{OC}^{L} + (\Delta y_{a,L} + \Delta y_{b,L} + \Delta y_{c,L}) \cos(L\theta)$$

$$Y_{OS}^{(L+1)} = Y_{OC}^{L} + (\Delta y_{a,L} + \Delta y_{b,L} + \Delta y_{c,L}) \sin(L\theta)$$

$$Y_{1C}^{(L+1)} = Y_{1C}^{L} + \Delta y_{a,L}\cos(L\theta) + \Delta y_{b,L}\cos(L+4)\theta + \Delta y_{c,L}\cos(L-4)\theta$$

$$Y_{1S}^{(L+1)} = Y_{1S}^{L} + \Delta y_{a,L}\sin(L\theta) + \Delta y_{b,L}\sin(L+4)\theta + \Delta y_{c,L}\cos(L-4)\theta$$

$$Y_{2C}^{(L+1)} = Y_{2C}^{L} + \Delta y_{a,L}\cos(L\theta) + \Delta y_{b,L}\cos(L-4)\theta + \Delta y_{c,L}\sin(L+4)\theta$$

$$Y_{2S}^{(L+1)} = Y_{2S}^{L} + \Delta y_{a,L}\sin(L\theta) + \Delta y_{b,L}\sin(L-4)\theta + \Delta y_{c,L}\cos(L+4)\theta$$

$$(5.76)$$

The recursive expressions in Equation (5.76) are off by an additional factor of 1/3 because of the 1/3 in Equation (5.70). This means the half-cycle SCDFT is off by 1/9 and the full-cycle SCDFT is off by 1/18. For impedance calculations the constant multipliers will cancel in the division. It is only necessary to consider them if there is a problem with overflow in the calculations or if metering quantities such as power or voltage are required.

5.5.2 Transient monitor

When the data window spans the instant of fault inception, as seen in Figure 5.1, the results obtained from almost any algorithm are unreliable. The difficulty is that the data contain both pre-fault samples and post-fault samples so that the fit of the estimate to the data is particularly bad. If the quality of the estimate could be determined, then the relay could be disabled during this transition. Using the least squares solution given by Equation (5.23), we can compute the sample values that would correspond to the estimate. If we denote these as the vector $\tilde{y}$ then

$$\tilde{y} = S\,\hat{Y} = S(S^{T}S)^{-1}S^{T}y \qquad (5.77)$$

The vector $\tilde{y}$ represents the reconstruction of the samples from the estimate $\hat{y}$, as shown in Figure 5.23. If $\tilde{y}$ is close to y then the estimate is trustworthy. If not,

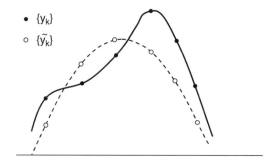

Figure 5.23 The actual samples and the reconstructed samples

then it is fair to assume that something such as the pre-fault/post-fault transition is involved. The difference can be formed as

$$r = \tilde{y} - y = [S(S^TS)^{-1}S^T - I]y \tag{5.78}$$

$$r = M y \tag{5.79}$$

The residual **r** can be thought of in much the same way as the observation residual in the Kalman filter. For the Fourier algorithms the matrix in Equation (5.78) can be simplified. With six Samples per half-cycle, for example

$$M = \frac{1}{3} \begin{bmatrix} 2 & -\sqrt{3}/2 & -1/2 & 0 & 1/2 & \sqrt{3}/2 \\ -\sqrt{3}/2 & 2 & -\sqrt{3}/2 & -1/2 & 0 & 1/2 \\ -1/2 & -\sqrt{3}/2 & 2 & -\sqrt{3}/2 & -1/2 & 0 \\ 0 & -1/2 & -\sqrt{3}/2 & 2 & -\sqrt{3}/2 & -1/2 \\ 1/2 & 0 & -1/2 & -\sqrt{3}/2 & 2 & -\sqrt{3}/2 \\ \sqrt{3}/2 & 1/2 & 0 & -1/2 & -\sqrt{3}/2 & 2 \end{bmatrix}$$

It can be shown (Problem 5.5) that the residuals r_k obey a recursive relationship similar to Equations (5.37) and (5.38). If we define a *transient monitor* function as the sum of the absolute values of the r_k, i.e.

$$t = \sum_{k=1}^{6} |r_k| \tag{5.80}$$

then t can be computed with much the same procedure as the fundamental frequency components or the symmetrical components. The t function is computed for each phase current. Figure 5.24 shows the transient monitor for the *a* phase current

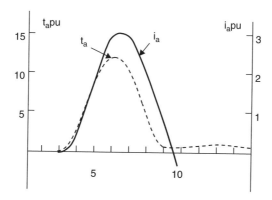

Figure 5.24   The transient monitor function for a phase a to ground fault

during a phase a to ground fault. It can be seen that the transient monitor behaves as desired. The value of t_a is large while the data window contains both pre-fault and post-fault data and drops when the window contains only post-fault data. If the transient monitor is large for any phase, the trip signal is inhibited. The transient monitor may pick up, i.e. exceed some threshold value, for transient events which are not faults. This is not troublesome since the transient monitor is only a control parameter which testifies to the purity of the data. However, the threshold value must be set high enough so that the random signals encountered in the post-fault currents do not inhibit tripping.

5.5.3   Speed reach considerations

If we examine the performance of the impedance like algorithms by taking the covariance of the estimate as a measure then Equation (5.22) implies a relationship between the length of the data window and the estimation error. For a measurement error covariance matrix which is a multiple of a unit matrix, $\mathbf{W} = \sigma_e^2\,\mathbf{I}$, the covariance of the estimation error is

$$
E\{(\hat{Y} - Y)(\hat{Y} - Y)^T\} = \sigma_\varepsilon^2
\begin{bmatrix}
\sum_{k=1}^{K} \cos^2(k\theta) & \sum_{k=1}^{K} \cos(k\theta)\sin(k\theta) \\
\sum_{k=1}^{K} \cos(k\theta)\sin(k\theta) & \sum_{k=1}^{K} \sin^2(k\theta)
\end{bmatrix}^{-1}
\tag{5.81}
$$

As in Figure 5.8, two quantities are being estimated, Y_c and Y_s. The diagonal entries in the inverse matrix in Equation (5.81) give the variances of each of the estimation errors while the off-diagonal entries give a cross covariance between the two. At multiples of a half-cycle the two estimation errors are independent. The various sums are

$$
\sum_{k=1}^{K} \cos^2(k\theta) = \frac{(K-1)}{2} + \frac{\cos^2 K\theta}{2} + \frac{1}{2}\sin 2K\theta\,\frac{\cos\theta}{\sin\theta}
$$

$$
\sum_{k=1}^{K} \sin^2 k\theta = \frac{(K+1)}{2} - \frac{\cos^2 K\theta}{2} + \frac{1}{2}\sin 2K\theta\,\frac{\cos\theta}{\sin\theta}
\tag{5.82}
$$

$$
\sum_{k=1}^{K} \cos(k\theta)\sin(k\theta) = \frac{1}{4}\sin 2K\theta + \frac{\sin^2 K\theta}{2}\,\frac{\cos\theta}{\sin\theta}
$$

For $K\theta$ a multiple of π the estimation errors are independent and have equal variances of $2\sigma_e^2/K$. The determinant of the matrix in Equation (5.81) is complicated in general but can be evaluated from the sums in Equation (5.82). If we add the two variances to obtain a variance for the phasor $\hat{Y}_c + j\hat{Y}_s$ we obtain

$$\sigma_{\hat{Y}}^2 = \frac{4\sigma_e^2}{K} \tag{5.83}$$

Before we can reach any firm conclusion about the behavior of the estimation error we must examine the term σ_e^2. To clarify the notation let us consider that $\mathbf{y}$ represents samples of voltage. If, motivated by Section 5.1, we imagine that the non-fundamental frequency signals that are being sampled to produce the error ε in Equation (5.18) are a wide-sense stationary random process (Section 3.7). Example 3.20, with $\omega_1 = 0$ and ω_2 very large) with a power density spectrum which is flat at a level S_v, i.e.

$$S(\omega) = S_v \text{ for all } \omega$$

then we will have to filter the noise with the anti-aliasing filter before sampling. If we assume an ideal filter with a cut-off frequency of

$$\omega_c = \frac{\pi\omega_o}{8} \tag{5.84}$$

one-half the sampling frequency where $\theta = \omega_0\Delta t$ then the noise before sampling has a spectrum

$$S_s(\omega) = S_v; \ |\omega| < \pi\omega_o/\theta$$

$$S_s(\omega) = 0; \ |\omega| > \pi\omega_o/\theta \tag{5.85}$$

where the sub s denotes sampling. From the definition of the correlation function (Equation (3.78) and the relationship between the correlation function and the spectrum (Equation (3.81)) we obtain

$$\sigma_e^2 = R_v(0) = \frac{1}{2\pi} \int_{-\pi\omega_o/\theta}^{\pi\omega_o/\theta} S_v \, d\omega = \frac{S_v\omega_o}{\theta} \tag{5.86}$$

or

$$\sigma_{\hat{V}}^2 = \frac{4\,S_v\omega_o}{K\theta} = \frac{4\,S_v}{T} \tag{5.87}$$

where T is the length of the window in seconds. If the impedance to the fault is computed as a ratio of estimated voltage to estimate current and we assume that

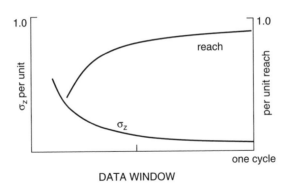

Figure 5.25   The variance of the impedance estimate and the reach versus the data window

the errors are small then the variance of the impedance error is given by

$$\sigma_{\hat{Z}}^2 = \sigma_{\hat{V}}^2 + \sigma_{\hat{I}}^2 = \frac{4(S_v + S_I)}{T} \tag{5.88}$$

Equation (5.88) is valid for T a multiple of a half-cycle of the fundamental frequency. The curve of the standard deviation (σ_z) of the impedance estimate obtained from Equation (5.81) is shown in Figure 5.25.

Several interesting observations can be made from Equation (5.88). The first concerns the sampling rate used in relaying algorithms. The sampling rate does not appear in Equation (5.88). The variance of the impedance estimate is independent of the sampling rate and is inverse to the length of the data window. The explanation lies with the anti-aliasing filter. At higher sampling frequencies the anti-aliasing filter has a larger bandwidth and each sample has a larger variance Equation (5.86). On the other hand, there are more samples in the same time period T and K is correspondingly larger (see Equation (5.83)). This effect will hold until the sampling frequency is so large that the assumption that the noise before filtering is white ($S(\omega) = S_v$ for all ω) is invalid. Since the noise is ultimately band-limited, for example, by the transducers, t here are sampling frequencies that are high enough to exceed the noise bandwidth. Laboratory and simulation results indicate that sampling frequencies in the kHz range would be required before this effect would be pronounced.[5]

There is an additional effect caused by the use of a non-ideal anti-aliasing filter. The actual anti-aliasing filter (Chapter 1) has a phase shift which translates into a delay. This delay, for well designed filters, is roughly one sample time. Lower sampling frequencies mean longer delays introduced by the anti-aliasing filter.

This analysis has been for the Fourier algorithms but is appropriate for all of the algorithms described in this chapter. A longer window will produce better estimates for all of the algorithm types. The counter scheme used for the differential-Equation algorithms is an attempt to create a longer window. Higher sampling rates (with the appropriate anti-aliasing filter) produce more noise per sample. In general, independent of sampling rate, the variance of the estimated fault location is inverse to the length of the data window.

The second curve in Figure 5.25 is also an important general result. If we accept that the estimated fault location is a random variable which has the correct mean but has a probability density with the variance given by Equation (5.88) then we must accept that there is a probability of false trip or failure to trip. The two situations are shown in Figure 5.26. In each case the density is drawn with its mean at the true fault location and the shaded area of the density corresponds to incorrect relay operation. If the density has *tails*, as shown in Figure 5.26, then there is no setting for the relay that can eliminate the failure to trip shown in Figure 5.26(a).

However, if we assume that almost all reasonable densities are concentrated within $\pm 2.5\sigma$ then we can conclude that the maximum relay setting that can safely be used is $1 - 2.5\sigma$. This would correspond to better than 99% confidence if the distribution were Gaussian and would guarantee no failures to trip if the density were triangular or uniform. The inverse time dependence of the variance of the estimates translates into the reach curve shown in Figure 5.25.

The implications of the speed-reach relationship are plain and are inherent in electromechanical relays. Close-in faults may be cleared quickly but faults near the zone-1 boundary require more processing. While the principle is physically appealing and is recognized in relaying practice,[21] it has an impact on digital schemes which use a fixed data window for ease of computation. For example, a fixed one half-cycle algorithm cannot clear a close-in fault as quickly as might be justified and cannot be set as near the zone-1 boundary as a longer window algorithm.

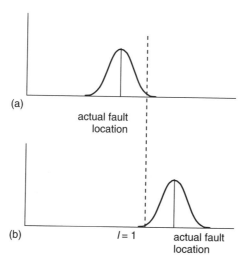

Figure 5.26 Distributions of estimated fault location. (a) Fault within zone-1. (b) Fault beyond zone-1

Evidently, what is needed is an adaptive scheme which adjusts the window length to the estimated fault location. An accumulator for this purpose has been suggested.[22]

5.5.4 A relaying program

As an example of the additional features that must be included in a complete line relaying program we will conclude with a description of the Symmetrical Component Distance Relaying Program (SCDR).[7] The flow chart for the program is shown in Figure 5.27. The program includes the following features:

1. Phase and Ground Distance Protection
2. Directional Comparison Carrier Blocking Scheme
3. Three-zone Stepped Dist ace for both Carrier and Time-Delay Backup
4. Local Breaker Failure Protection
5. High Speed Reclosing Initiation
6. Automatic Reclosing with Synchronism Check
7. Sequence of Events Recording
8. Fault Classification
9. Single-Phase-Trip Output
10. Memory Voltage for Three-Phase Bus Fault
11. Directional Inverse Time Backup for Ground Faults using Ground Current.

The SCDFT, K algorithm, and transient monitor algorithm have been described earlier in this section. The circuit-breaker failure flag (CBF) is set high following a trip command to the line breakers.

The CBF routine initiates a timer and upon expiration of the time interval checks the line currents. If the currents are not zero, a CBF trip signal is sent to the appropriate breakers. The value of k computed by the K routine is checked to see if it lies in various zones of the relay. A value of k lying in zone-1 or in zone-3 with no carrier received, leads to High Speed Reclose (HSR) output and breaker trip output from each is provided when the time exceeds the appropriate time delay. The CARRIER START routine is entered and the local carrier transmitter is started if k is in the carrier start zone.

The Auto Reclose Flag is set high following the BREAKER TRIP routine. The AR routine which is entered when this flag is high maintains timers and upon their expiration checks the phase angle between the bus voltage of phase a and the positive sequence line voltage. If these two voltage are within a preset limit, Auto-Reclose is initiated. The Single-Phase Trip feature of the Trip routine uses the phase angle between the positive and negative sequence currents computed by the SCDFT. The entire program is repeated at each sample time (12 times a cycle of the nominal fundamental frequency).

Several additional features of the k calculations are addressed in the problems. The correct k is obtained even if one of the voltages is missing due to a blown fuse in a CVT (Problem 5.8). The k maintains its directionality for unbalance

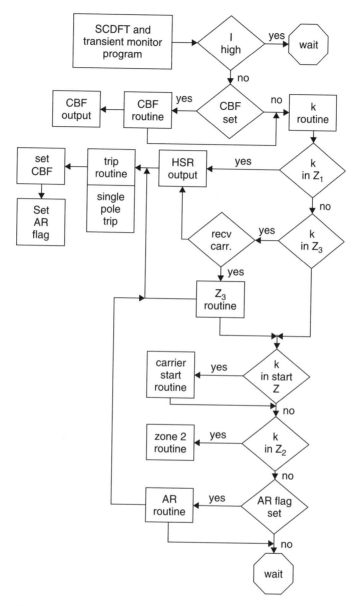

Figure 5.27   The flow chart for the symmetrical component distance relay

faults even if the fault is a so-called *bolted fault* producing zero bus voltage (Problem 5.9).

Except for the fact that this program uses the symmetrical component Equation (5.75) for distance calculations, this could be a relaying program using any other method for computing the distance to the fault. The complexity of the program beyond the distance calculation should be borne in mind, as often the impact of an improvement in the distance calculation technique is completely masked by the larger relaying programs in which the algorithm must be imbedded.

5.6 Newer analytic techniques

Some of the techniques presented in Chapter 4 have been proposed in areas beyond line protection, such as windowing for PMUs when the system frequency is off nominal. Wavelets have been suggested for high impedance fault detection, transformer protection, fault classification, fault location, adaptive reclosing, and rotating machinery protection. As mentioned in the preface, the case for the application of artificial intelligence to conventional first zone protection has not really been made.

5.6.1 Wavelet applications

An early wavelet application that took advantage of the wavelet's superior time frequency resolution was in dealing with the complicated waveforms that are generated by resonant grounding.[23] It is common in Europe to insert a Petersen coil in the connection of the neutral to ground. When properly tuned this eliminates arcing grounds. For a single phase fault the signal measured by a zero sequence device contains significant transient components and only small amounts of the fundamental frequency contribution, however. The signals are non-stationary short-duration transients and are difficult to deal with in a Fourier environment. With a carefully chosen Mother Wavelet it is possible to use the wavelet transform magnitude and phase to set thresholds to distinguish faults from non faults.

Wavelets have been suggested in the environment of the traveling wave relays of Chapter 9.[24] The issue is determining fault location from observations of the time of arrival of traveling waves caused by a fault on a transmission line. In Chapter 9, a discriminant function is presented to solve this problem. Correlation techniques have also been suggested for this purpose. Because the wavelet transform is capable of localizing in both time and frequency it is an attractive option especially with synchronized sampling at both ends of the line.

Both neural nets and decision trees have been proposed to learn the difference between a fault and inrush in transformer protection.[25] The conventional harmonic approach to this problem is presented in Chapter 6. Even the combination of wavelet analysis and decision trees has been proposed.[26] The wavelet coefficients are the variables in the decision tree construction.

5.6.2 Agent applications

Agents have been proposed for the protection of tapped transmission lines and for a variety of backup protection applications.[27,28] The issues of communication traffic patterns and the connection with IEC 61850 and message structure and strategies for such systems have been studied in simulation.[29]

5.7 Protection of series compensated lines

One of the earliest field installations of computer relays dealt with the problem of protecting a series compensated transmission line.[30] In order to improve the power transfer capability of a long distance transmission line, it is often necessary to insert a three-phase capacitor bank in series with the transmission line. Usually the series capacitors are inserted at one end of the line, but occasionally they may be installed in the middle, or even in more than one location. A compensating capacitor having a reactance equal to the positive sequence reactance of the transmission line is said to provide 100% compensation, while a capacitor with a smaller reactace, i.e. with a larger value of capacitance) provides a proportionally smaller compensation. In most cases, the compensation is smaller than 100%. Consider the compensated transmission line shown in Figure 5.28(a). As long as the capacitors are in service, the impedance seen by distance relays at terminal A varies along the segmented line ABCD as shown in Figure 5.28(b). It is clear that an impedance measured by the relay at A may not have a one-to-one correspondence with the distance to the fault, particularly if the distance is primarily determined by the fault reactance X_f. In some cases it may be that the resistance of the measured impedance is sufficiently well defined to make the distance calculation based upon the complex impedance a practical proposition.

The actual fault conditions are likely to be more complex, as the protective gaps or lightning arresters across the capacitors may flash over for some transmission line faults. Should this happen, the distance measurement would change suddenly when the capacitor protection system operates. This is likely to be a single-phase

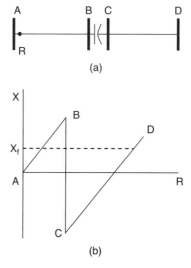

Figure 5.28 R-X diagram for a series compensated line. (a) One-line diagram. (b) R-X diagram

phenomenon, and consequently the fault calculation is further complicated by the presence of a simultaneous unbalance: the original fault, and the short-circuit across one phase of the compensating capacitor. The distance calculated from the terminal voltage and current is now a very complex function of the distance to the fault, the amount of series compensation, and the placement of the series compensation. It should be clear from this discussion that the determination of fault location from a ratio of fundamental frequency voltage and current at the terminals of a series compensated transmission line is not a simple matter.

Even the calculation of voltage and current phasors is fraught with difficulty in this case. As mentioned earlier, the usual case of less than 100% series compensation requires that capacitors greater than those required for 100% compensation be connected in series with the transmission line. These capacitors form resonant circuits with the reactance of the transmission line and of the source system. Consequently the resonance frequencies of these circuits are smaller than the fundamental power frequency. These extraneous frequencies – which are seldom sub-multiples of the power frequency – cause considerable error in small window phasor calculations. It is necessary to devise algorithms which reject these low frequencies, as well as those above the fundamental frequency. The reference cited earlier used first and second derivatives of the current and voltage signals to reject the low frequency components. Consider an input $y(t)$ having two frequency components:

$$y(t) = a_1 \sin \omega_1 t + a_0 \sin (\omega_0 t + \varphi) \tag{5.89}$$

where ω_0 is the power frequency, and ω_1 is a frequency considerably smaller than ω_0. Calculating the first and second derivatives:

$$\dot{y}(t) = a_1 \omega_1 \cos \omega_1 t + a_0 \omega_0 \cos (\omega_0 t + \varphi) \tag{5.90}$$

$$\ddot{y}(t) = -a_1 \omega_1^2 \sin \omega_1 t - a_0 \omega_0^2 \sin (\omega_0 t + \varphi) \tag{5.91}$$

If ω_1 is much smaller than ω_0, the first term on the right hand sides of Equations (5.90) and (5.91) can be neglected, and one can solve for the magnitude and phase angle of the fundamental frequency component:

$$a_0^2 = \dot{y}/\omega_0)^2 + (\ddot{y}/\omega_0^2)^2 \tag{5.92}$$

$$\varphi + \omega_0 t = \arctan (-\ddot{y}/\omega_0 \dot{y}) \tag{5.93}$$

One could design a least squares solution in the spirit of Chapter 3 with similar assumptions about the two frequencies. However, as is to be expected, the resulting algorithms are far more sensitive to noise

Estimation of fault location from terminal voltages and currents of a series compensated transmission line remains a difficult problem. As in traditional relaying, phase-comparison relaying and longitudinal differential protection are the protection schemes of choice for such lines.

5.8 Summary

In this chapter we have examined line relaying algorithms. The interested reader is referred to other books on the subject.[31,32] We have seen that the fundamental limitation of all of the various algorithms is the presence of unpredictable non-fundamental frequency signals in the voltage and current waveforms just after fault inception. Whether the algorithm is attempting to estimate the fundamental frequency components of voltage and current to estimate the impedance of using a series R-L model of the line, these unmodeled signals cause errors in the estimated fault location. The role of the exponential offset in the fault current is an important consideration in differentiating between algorithms. If the offset is removed outside to the relaying algorithm by analog filtering or a separate subroutine, then the Fourier type algorithms offer considerable advantages in terms of simplicity and performance. The differential-equation algorithms do not require that the offset be removed but do have some performance limitations for long high-voltage lines if the system structure at the source bus is complicated. If the error terms in the measured current and voltage are assumed to have a significant variation in their statistical description during the relaying interval, the Kalman filtering approach may be indicated. The increased computation burden must be considered along with problem of the need for a more detailed error model.

Additional features of relaying algorithms such as their ability to determine the fault type should also be considered. The use of Clarke or Symmetrical Components provides a technique for determining fault type for the algorithms presented. The consequence of mis-classification of the fault is an additional consideration in algorithm selection.

Finally, there is an inherent speed-reach limitation in all distance relaying. Given the unpredictable nature of the non-fundamental frequency components of the post-fault current ad voltage, close-in faults may be cleared confidently with short window algorithms but the remove reach of the relay requires a longer data window. The distant reach of fixed window digital relaying algorithms must be set with these limitations in mind. An adaptive speed-reach characteristic is an important feature of any relaying program.

Problems

5.1 Determine a value of c_1 in Equation (5.6) so that $Y_c = 0$ for the third harmonic.

5.2 Work out the algorithm suggested by Equation (5.23) if the signal set includes the three terms $s_1(t) = \cos \omega_o t$, $s_2(t) = \sin \omega_o t$ and $s_3(t) = e^{-(R/L)t}$.

5.3 Develop a recursive form of a two-cycle Fourier algorithm with a sampling rate of four samples per cycle.

5.4 Compare the answer to Problem 5.2 with Equation (5.67).

5.5 Show that the components of the residual vector r in Equation (5.79) obey a recursive relationship similar to Equations (5.37) and (5.38).

5.6 Derive the Symmetrical Component Distance Relay equations.

5.7 Derive distance calculation equations for two lines with mutual coupling.

5.8 Show that the SCDR produces the correct distance calculation even if one of the voltages is due to a blown fuse.

5.9 Show that the SCDR maintains directionality for all unbalanced faults even when the fault is a bolted fault.

5.10 Derive the response of a *b-c* relay
  (a) to an a-b fault
  (b) to a b-g fault.

5.11 Calculate the voltage changes in phase *a, b, c* at the relay location of a line fed from a source as a function of the Source to Line Impedance Ratio (SIR). Assume suitable X_m/X_s and neglect R and C.

5.12 Assume a π-section single-phase model of a transmission line. Calculate the voltage and current transients for a line-end fault. Determine the error in the differential-equation algorithm if the source is purely inductive. Assume suitable values for L and C for (a) a long line and (b) a short line.

References

[1] Gilbert, J.C., Udren, E.A. and Sackin, M. (1977) Evaluation of algorithms for computer relaying, IEEE Publication no. 77CH1193 PWR, Paper no. A77-520-0, IEEE PES Summer Meeting, Mexico City, pp. 1–8.
[2] B.J. Mann, and I.F. Morrison (1971) Digital calculation of impedance for transmission line protection, IEEE Trans. on PAS, vol. 90, no. 1, pp. 270–279.
[3] Gilchrist, G.B., Rockefeller G.D. and Udren, E.A. (1972) High-speed distance relaying using a digital computer, Part I: System description, IEEE Trans. on PAS, vol. 91, no. 3, pp. 1235–1243.
[4] Thorp, J.S., Phadke, A.G., Horowitz S.H. and Beehler, J.E. (1979) Limits to impedance relaying, IEEE Trans. on PAS, vol. 98, no. 1, pp. 246–260.
[5] Sachdev M.S. and Baribeau, M.A. (1979) A Digital computer relay for impedance protection of transmission lines, Trans. of Engineering and Operating Division Canadian Electrical Association, vol. 18, Part 3, no. 79-SP-158, pp. 1–5.

[6] Luckett, R.G., Munday, P.J. and Murray, B.E. (1975) A substation-based computer for control and protection, Developments in Power System Protection, IEE Conference Publication No. 125, pp. 252–260.

[7] Phadke, A.G., Hlibka, T., Ibrahim, M. and Adamiak, M.G. (1979) A microprocessor based symmetrical component distance relay, Proceedings of PICA, Cleveland.

[8] McInnes, A.D. and Morrison, I.F. Real time calculations of resistance and reactance for transmission line protection by digital computer (1970) Elec. Eng. Trans. IE, Australia, vol. EE7, no. 1, pp. 16–23.

[9] Ranjbar, A.M. and Cory, B.J. (1975) An improved method for the digital protection of high voltage transmission lines, *IEEE Trans. on PAS*, vol. 94, no. 2, pp. 544–550.

[10] Breingan, W.D. Chen, M.M. and Gallen, T.F. (1979) The laboratory investigation of a digital system for the protection of transmission lines, IEEE Trans. on PAS, vol. 98, no. 2, pp. 350–368.

[11] Smolinski, W.J. (1980) An algorithm for digital impedance calculation using a single pi section, IEEE Trans. on PAS, vol. 98, no. 5, pp. 1546–1551, and vol. 99, no. 6, pp. 2251–2252.

[12] Girgis, A.A. and Brown, R.G. (1981) Application of Kalman filtering in computer relaying, IEEE Trans. on PAS, vol. 100, no. 7, pp. 3387–3397.

[13] Girgis A.A. and Brown, R.G. (1983) Modeling of fault-induced noise signals for computer relaying applications, IEEE Trans. on PAS, vol. 102, no. 9, pp. 2834–2841.

[14] Sachdev, M.S., Wood H.C. and Johnson, N.G. (1985) Kalman filtering applied to power system measurements for relaying, IEEE Trans. on PAS, vol. 104, no. 12, pp. 3565–3573.

[15] Swift, G.W. (1979) The spectra of fault induced transients, IEEE Trans. on PAS, vol. 98, no. 3, pp. 940–947.

[16] Phadke, A.G., Hlibka, T., Adamiak, M.G., Ibrahim, M. and Thorp, J.S. (1981) A microcomputer based ultra-high speed distance relay: Field tests, IEEE Trans. on PAS, vol. 100, no. 4, pp. 2026–2036.

[17] Centeno, V. (1988) Mimic circuit simulation in real time, MS Thesis, Virginia Tech.

[18] Mann B.J. and Morrison, I.F. (1971) Relaying a three-phase transmission line with a digital computer, IEEE Trans. on PAS, vol. 90, no. 2, pp. 742–750.

[19] Clarke, E. (1943) *Circuit Analysis of A-C Power Systems*, vol. I, John Wiley & Sons, Inc. New York.

[20] Phadke, A.G., Hlibka T. and Ibrahim, M. (1977) Fundamental basis for distance relaying with symmetrical components, IEEE Trans. on PAS, vol. 96, no. 2, pp. 635–646.

[21] Andrichak J.G. and Wilkinson, S.B. (1976) Considerations of speed, dependability and security in high speed pilot relaying schemes, Minnesota Power Systems Conference, October 1976.

[22] Udren E.A. and Sackin, M. (1980) Relaying Features of an Integrated Microprocessor-Based Substation Control and Protection System, IEE Conference Publication no. 185, London, U.K., April 1980, pp. 97–101.

[23] Chaari, O., Meunier, M. and Brouaye, F. (1996) Wavelets: a new tool for the resonant grounded power distribution systems relay, IEEE Transactions in Power Delivery, vol. 11, no. 3, pp. 1301–1308.

[24] Magnago F.H. and Abur, (1998) A. Fault location using wavelets, IEEE Trans on Power Delivery, vol. 13, no. 4, pp. 1475–1480.

[25] Nagpal, M., Sachdev, M.S., Kao, N., Wedephol, L.M. (1995) Using a Neural Network for transformer protection, International Conference on Energy Management and Power Delivery, Proceedings of EMPD '95, pp. 674–679.

[26] Sheng, Y. and Rovnyak, S.M. (2002) Decision trees and wavelet analysis for power transformer protection, IEEE Trans. on Power Delivery, vol. 17, no. 2, pp. 429–433.

[27] Coury, D.V., Thorp, J.S., Hopkinson, K.H. and Birman, K.P. (2002) An agent based current differential relay for use with a utility Internet, IEEE Trans. on Power Delivery, vol. 17, no.1, pp. 47–53.

[28] Giovanini, K. Hopkinson, R, Coury, D.V. and Thorp, J.S. (2006) A primary and backup cooperative protection system based on wide area agents, IEEE Trans. on Power Delivery, vol. 21, no. 3, pp. 1222–1230.

[29] Tong, X. Wang, X. and Ding, L. (2008) Study of information model for wide-area backup protection agent in substation based IEC 61850, Third International Conference on Electric Utility Deregulation and Restructuring and Power Technologies, 2008. Nanjing, China, pp. 2212–2216.

[30] Rockefeller, G.D. and Udren, E.A. (1972) High-speed distance relaying using a digital computer, Part II: Test results, IEEE Trans. on PAS, vol. 91, no. 3, pp. 1244–1252.

[31] Johns A.T. and Salman, S.K. (1995) *Digital Protection for Power Systems*, Peregrinus, Ltd.

[32] Ziegler, B. (1999) *Numerical Distance Protection*, Siemens.

6

Protection of transformers, machines and buses

6.1 Introduction

This chapter is concerned with algorithms for the protection of power transformers, generators and buses. All of the devices can be protected with algorithms based upon ideas of differential protection since measurements at all of the terminals can be made available to the algorithm. Each type of algorithm, however, must deal with effects which tend to confuse the percentage differential characteristic. Algorithms for the protection of power transformers must be designed to operate correctly in the presence of magnetizing inrush (which appears to be an internal fault to the percentage differential characteristic) and in the presence of over excitation, which has a similar effect. Saturation of the current transformers is an additional issue in power transformer protection and is a fundamental issue in bus protection. Generator protection usually consists of many tasks that include control functions as well. Differential protection of generators is usually a straightforward task, since the current transformers used are matched and sized carefully to produce very little differential current for external faults. Motor protection also incorporates many control tasks.

These issues and the conventional solutions to these problems are described in Chapter 2. In some cases, such as the use of harmonic restraint for dealing with inrush and over-excitation (Section 2.4), there is an almost direct digital implementation of a conventional relaying practice. There are, however, subtle differences in the way analog filters obtain the harmonic content of signals as opposed to digital techniques. In other cases of conventional relaying practice, such as high impedance bus protection (Section 2.5), there is no digital implementation and new digital solutions must be found. For generators, phasor calculation of

Computer Relaying for Power Systems 2e   by A. G. Phadke and J. S. Thorp
© 2009 John Wiley & Sons, Ltd

currents, voltages and frequency measurements are sufficient to satisfy all protection needs.

In an integrated digital substation protection system, new solutions to some of these problems are made possible by an increase in the data that can be used in decision making. Conventional power transformer protection uses only current measurements. In an integrated substation it is assumed that all the samples of voltages and currents will be synchronized so that data can be shared, both for backup and for improved decision making. The use of voltage measurements is then possible in transformer protection. In the next section a number of different power transformer algorithms will be described, some of which closely follow conventional approaches and some of which are unconventional in terms of their use of voltage measurements. The reader is referred to Section 2.4 for the background material on transformer protection.

6.2   Power transformer algorithms

Some version of the percentage differential characteristic is a part of almost all proposed transformer algorithms. The only addition that should be made to Section 2.4 and Figure 2.11 is that for multi-winding transformers there are a number of restraint currents. If we consider a three-winding transformer as shown in Figure 6.1 with the current polarities as shown, the trip current is given by

$$I_T = I_1 + I_2 + I_3$$

while there are two restraining currents given by

$$I_{R1} = I_1 - I_2 - I_3$$

and

$$I_{R2} = -I_1 + I_2 - I_3$$

The second restraint current is necessary to protect the transformer operating with the primary breaker open. Although much of the original algorithm development

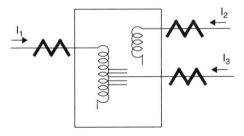

Figure 6.1   A three-winding transformer

has been on a single-phase basis, the ultimate application is on a three-phase device. Thus, there would be three trip currents (one per phase) and six restraint currents (two per phase) for a three-phase three-winding transformer.

While the percentage differential characteristic of Figure 2.11 could be applied to each phase (checking each of the restraint currents against the trip current) on a per sample basis, it is clear that there are error terms in the samples (as in line relaying) and that some filtering would improve performance. More importantly, the percentage differential characteristic must be inhibited during periods of magnetizing inrush or over-excitation. The first class of transformer algorithms includes those which form a restraint for the percentage differential from the current measurements themselves.

6.2.1 Current derived restraints

As pointed out in Section 2.4, the presence of second harmonic in the transformer inrush current and the presence of fifth harmonic in the current that flows under conditions of over-excitation lead to the principle of harmonic restraint. In analog relays filters are used to obtain some combination of non-fundamental frequency components. When the output of these filters is high the relay is restrained. Early digital versions of the harmonic restraint-percent-differential relay differ in their technique for obtaining the harmonic content of the currents. Techniques based on wave shape identification,[1] on the use of recursive band-pass digital filters,[2] on cross correlation with sinusoids or square waves,[3] on finite impulse response digital filters,[4] on Walsh type transforms,[5] least square curve fitting,[6] and the discrete Fourier transform[7] have been proposed. The wave shape identification is based on the observation that under inrush conditions the peaks of the current wave are either closer together or farther apart than is normal. The remaining algorithms are all attempts to obtain simple computational techniques for determining the harmonic content of the currents. The concern with computational efficiency is motivated by the large number of currents that would be involved for a three-phase multi-winding transformer. As was pointed out in Chapter 4, the discrete Fourier transform produces optimal estimates if a constant error covariance is assumed. The recursive form of the DFT for fundamental and harmonic frequencies is particularly simple with an appropriate sampling rate. These recursive DFT calculations were found to represent the best compromise between speed of calculation and accuracy of the results.[8]

The recursive form of the full-cycle Fourier calculation of the nth harmonic (in complex form – see Equation (4.34)) for samples ending at sample L is given by

$$Y_n^{(L)} = Y_n^{(L-1)} + \frac{2}{K}[y_L - y_{L-K}]e^{-jnL\theta} \qquad (6.1)$$

where $K(\theta) = 2\pi$. The difference between Equations (6.1) and (4.34) is the n in the exponent and the scaling of 2/K that was omitted in Equation (4.34). That is, the

Table 6.1 One cycle of coefficients for the fundamental, the second harmonic, and the fifth harmonic with a sampling frequency of 12 samples per cycle

Harmonic	1	1	1	1	2	2	2	2	5	5	5	5
Real	X	X			X	X			X	X		
Imaginary			X	X			X	X			X	X
Channel Gain												
1.0	X		X		X		X		X		X	
0.866		X		X		X		X		X		X
n												
1	1	0	0	0	1	0	0	0	1	0	0	0
2	0	1	1/2	0	1/2	0	0	1	0	-1	1/2	0
3	1/2	0	0	1	$-1/2$	0	0	1	1/2	0	0	-1
4	0	0	1	0	-1	0	0	0	0	0	1	0
5	$-1/2$	0	0	1	$-1/2$	0	0	-1	$-1/2$	0	0	-1
6	0	-1	1/2	0	1/2	0	0	-1	0	1	1/2	0
7	-1	0	0	0	1	0	0	0	-1	0	0	0
8	0	-1	$-1/2$	0	1/2	0	0	1	0	1	$-1/2$	0
9	$-1/2$	0	0	-1	$-1/2$	0	0	1	$-1/2$	0	0	-1
10	0	0	-1	0	-1	0	0	0	0	0	-1	0
11	1/2	0	0	-1	$-1/2$	0	0	-1	1/2	0	0	1
12	0	1	$-1/2$	0	1/2	0	0	-1	0	-1	$-1/2$	0

correction term for the nth harmonic is the difference between the newest sample and the oldest (one cycle old) multiplied by an angle that rotates more rapidly for the higher harmonics.

At a sampling rate of 12 samples per cycle the only irrational number involved in the update is $\sqrt{3}/2$. If this multiplication is accomplished externally by a voltage divider or in software, as in Example 4.2, then a table of coefficients for the multiplications in Equation (6.1) can be given in terms of the two channels of data, as shown in Table 6.1.

For a three-winding transformer the fundamental, second, and fifth harmonic must be computed for each of nine phase currents. There is evidence that a secure restraint function for a three-phase transformer can be formed by combining the harmonics from all three phases.[9] A typical algorithm might then form two harmonic restraints: one the sum of the three second harmonic magnitudes, and the second the sum of the three fifth harmonic magnitudes. Denoting these sums as I_{H2} and I_{H5}, the relay is restrained for a given phase if

$$|I_T| < \alpha$$

or

$$|I_T| < \beta |I_{R1}|$$

or

$$|I_T| < \beta |I_{R2}| \qquad\qquad (6.2)$$

or

$$|I_{H2}| < \gamma |I_T|$$

or

$$|I_{H5}| < \delta |I_T|$$

where α is the threshold value in Figure 2.11, β is the slope of the percentage differential characteristic and γ and δ are the percent harmonic restraints. The currents in Equation (6.2) are the full cycle phasor values computed from Equation (6.1). It should be noted that the factors affecting the choice of these parameters include not only the transformer type and steel but also the design of the anti-aliasing filters. The need for a fifth harmonic implies a sampling frequency of at least 12 samples per cycle. At that rate, the fifth harmonic is quite close to the anti-aliasing filter cut-off frequency and may be attenuated by the filter. The parameter γ must be selected with this effect in mind.

As an example of this type of algorithm, the response of the various restraint currents for an internal a to ground primary fault are shown in Figure 6.2. The fault occurred approximately at sample 14. Prior to sample 17, before any phase had developed a trip signal, the second and fifth harmonic restraints were effective. At sample 17 the phase a differential indicated a trip but the relay was restrained by the harmonics. The fifth harmonic drops out at sample 22 and the second harmonic restraint drops out at sample 27. The trip signal would be issued at sample 27, one cycle after fault initiation. In general the harmonics develop more rapidly than the

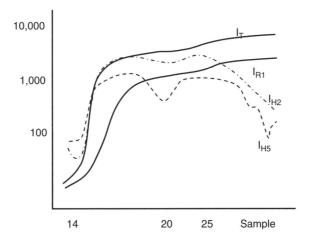

Figure 6.2   The response of the restraint and trip currents to an internal a-ground fault

fundamental during a transient so that the relay is restrained for approximately a full cycle by the harmonic restraints. In effect the harmonic restraint acts much like the transient monitor of Chapter 5.

6.2.2 Voltage based restraints

As pointed out in the Introduction, it is reasonable in an integrated substation protection system to assume that bus voltage measurements would be available for transformer protection. While the requirement of additional voltage measurements would increase the cost of a stand-alone transformer protection unit, such voltage measurements may be obtained quite inexpensively in an integrated system. In fact, an early solution to the inrush problem used voltage measurements to restrain a percentage differential.[10] The so-called 'tripping suppressor' used a voltage relay to suppress the tripping function if the voltage was high. In its early analog form, the 'tripping suppressor' was found to be slower than harmonic restraint devices. Since the harmonic restraint algorithm is essentially a one cycle relay and since the short window line protection algorithms can compute voltage phasors in as little as a quarter of a cycle, it has been suggested that a digital 'tripping suppressor' may be faster than the digital harmonic restraint algorithm.[11]

The proposed algorithm used a one-half cycle window for the calculation of fundamental frequency components of trip and restraint currents and the primary voltage for each phase. It was determined that it was necessary to include a transient monitor (Section 4.5) for the voltage signals. Thus for a three-winding transformer, one trip and two restraint currents, the primary voltage and the transient monitor value must be updated for each phase. This total of 27 updates of the form of Equations (4.37) and (4.38) (counting real and imaginary parts for the voltage and current phasors) compares with 30 such simple updates for the harmonic restraint algorithm (the fundamental of the trip and two restraint currents and the second and fifth harmonic of one of the restraint currents for each phase). The saving is simply that t_V is real rather than complex.

The relay is restrained for a given phase if

$$|I_T| < \alpha$$

or

$$|I_T| < \beta |I_{R1}|$$

or

$$|I_T| < \beta |I_{R2}| \tag{6.3}$$

or

$$|V| > \sigma$$

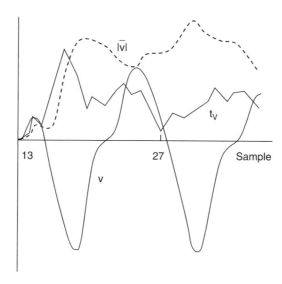

Figure 6.3   Phase c voltage quantities for the voltage restraint algorithm – inrush

or

$$|t_V| > \rho$$

The quantities in Equation (6.3) are now half cycle phasor results rather than the full cycle quantities of Equation (6.2). The first three inequalities in Equation (6.3) produce the percentage differential characteristic while the last two restrain if the voltage is high or the transient monitor indicates that the voltage phasor is unreliable. The transient monitor is necessary since the voltage waveform can be so distorted that the half cycle phasor for the voltage fails to be large enough. The voltage quantities for phase c for an inrush case are shown in Figure 6.3. The serious voltage distortion is due to the inrush. The inrush is sufficient to produce a current differential trip in phases a and c. For samples 17–23 only the transient monitor restrains phase a, while at sample 27 the monitor drops but the high voltage restrains phase c.

The voltage restraint algorithm is potentially faster than the harmonic restraint algorithm since it is based on a half cycle computation. It is also somewhat less complicated to implement on a microprocessor.

6.2.3   Flux restraint

There are other possible uses of measured voltages. It has been suggested that, using a linear model of the transformer, the measured currents could be used to compute the terminal voltages.[12] The comparison of the computed voltages with the measured voltages would then give an indication of whether the linear (unsaturated) model was correct or whether saturation had begun. Another possibility is to use the measured voltages and currents to determine the internal flux of the transformer.[13]

Neglecting the resistance of the winding, the voltage at the terminals of a transformer winding, $v(t)$, the current through that winding, $i(t)$, and the flux linkage, $\Lambda(t)$ of the transformer are related

$$v(t) - L\frac{di(t)}{dt} = \frac{d\Lambda(t)}{dt} \tag{6.4}$$

where L is the leakage inductance of the winding. If Equation (6.4) is integrated between sample times t_1 and t_2

$$\Lambda(t_2) - \Lambda(t_1) = \int_{t_1}^{t_2} v(t)\,dt - L[i(t_2) - i(t_1)]$$

Using trapezoidal integration as in Equations (4.45) and (4.46) and equally spaced samples, we can write

$$\Lambda(t_2) = \Lambda(t_1) + \frac{\Delta t}{2}[v(t_2) + v(t_1)] - L[i(t_2) - i(t_1)]$$

or with subscript k denoting the kth.

$$\Lambda_{k+1} = \Lambda_k + \frac{\Delta t}{2}[v_{k+1} + v_k] - L[i_{k+1} - i_k] \tag{6.5}$$

The current in Equation (6.5) is the trip current $I_1 + I_2 + I_3$ for each phase of the three-winding transformer. The voltage in Equation (6.5) is the winding voltage. In the case of delta windings, the line–line voltages would have to be obtained from the phase to neutral voltages. Similarly, if the transformer were grounded through an impedance, it would be necessary to account for the neutral potential. Equation (6.5) permits the computation of the mutual flux linkage from measured values of currents and voltages. If the initial flux linkage were known, then at each sample it would be possible to track the flux-current plot of the transformer. The $i - \Lambda$ plane is shown in Figure 6.4. The open circuit magnetizing curve of the transformer is shown along with the flux-current curve that would be associated with an internal fault. It can be seen that the flux-current characteristic can provide an effective restraint function. If the flux computed from Equation (6.5) and the measured current (i_k, Λ_k), lie on the open circuit magnetizing curve, then the relay should be restrained. There are distinct regions in the (i, Λ) plane corresponding to fault or no-fault conditions as shown. It should be noted that there are no phasor calculations involved and that the percentage differential and the flux restraint are

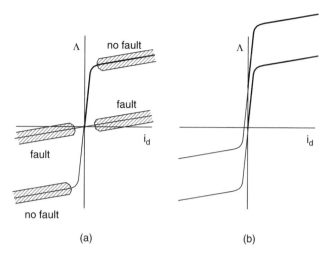

Figure 6.4 The flux-current plane. (a) Fault and no-fault regions. (b) Effect of unknown remnant flux

instantaneous. While the simplicity of computation is appealing, some averaging is clearly called for to provide security. In addition there is a problem in assuming the initial flux linkage is known. Figure 6.4(b) shows the effect of the unknown initial flux. Note that it is not possible to discriminate between fault and no-fault situations in the (i, Λ) plane with an unknown initial flux.

   Both problems can be resolved by considering the slope of the flux curve. From Figure 6.4(b) it can be seen that the open circuit magnetizing is characterized by different slopes in the (i, Λ) plane. Using Equation (6.5), we can form an expression for the slope (dΛ/di) as

$$\left(\frac{d\Lambda}{di}\right)_k = \frac{\Lambda_k - \Lambda_{k-1}}{i_k - i_{k-1}} = \frac{\Delta t}{2}\left[\frac{e_k + e_{k-1}}{i_k - i_{k-1}}\right] - L \tag{6.6}$$

Operation in the unsaturated region of the open circuit magnetizing curve Figure 6.4 produces a large value for the slope (dΛ/di) while the fault or no-fault (saturated) regions have much smaller slopes. It is not possible to distinguish between the fault or no-fault regions on the basis of the slope but it is not necessary to do so. During an internal fault the (dΛ/di) samples remain small (the fault region of Figure 6.4) continuously. During inrush, on the other hand, the (dΛ/di) samples alternate between large and small values as the magnetizing curve is traced. Thus if we define a positive restraint index k_r, as follows

$$k_r = k_r + 1$$

if the current differential indicates trip

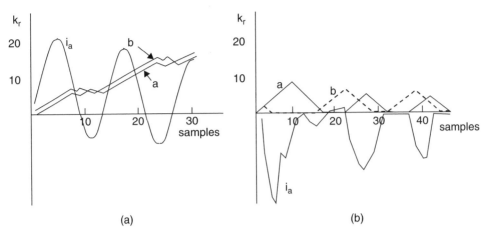

Figure 6.5 The behavior of the restraint index. (a) An internal a-b fault. (b) Inrush produced following the removal of an external three-phase fault

and if

$$(d\Lambda/di)_k > \zeta$$

$$k_r = \begin{cases} k_r - 1 & \text{if } k_r > 0 \text{ and } (d\Lambda/di)_k > \zeta \\ k_r & \text{if } k_r = 0 \end{cases} \tag{6.7}$$

then k_r will increase whenever the current differential shows a trip and the flux is not on the steep part of the magnetizing curve. In fact, the index grows almost monotonically for internal faults but shows a saw-tooth behavior for non-fault cases. The threshold value ζ simply separates the high and low slopes of the magnetizing curve.

The behavior of the restraint index for a phase a to b internal fault along with the phase a differential current are shown in Figure 6.5(a). The hesitation in the growth of k_r for phases a and b indicates that there were points on the fault current waveform when the transformer was saturated. The restraint index for phases a and b for an inrush condition along with the phase a are shown in Figure 6.5(b). The maximum value attained by k_r is six. For 21 cases reported in reference 13, six was the largest value ever encountered. Figure 6.5 is obtained for a sampling rate of 12 samples per cycle. It is clear that the maximum value of k_r depends on the sampling rate. The trip decision would be made when k_r exceeded some threshold k_{rrmax}. The choice k_{rrmax} of illustrates the familiar conflict between the speed and security of any relay. If k_{rrmax} is small the relay will be fast but will have some chance of a false trip. As k_{rrmax} is made larger the relay becomes more secure but slower.

With a multi-winding transformer it is possible that the transformer might be energized from any of its windings. The voltage required for calculating the flux must then be obtained from the appropriate bus. Again, in an integrated substation it is assumed that this voltage could be made available from the module which is protecting equipment connected to the energizing bus.

A similar algorithm for power transformer protection which also uses measured voltages and currents has been proposed[14]. Using a description of a three-winding transformer in terms of the reciprocal inductance matrix Γ, the terminal description of the transformer can be written as

$$
\begin{bmatrix} \Delta I_1 \\ \Delta I_2 \\ \Delta I_3 \end{bmatrix} = \begin{bmatrix} \Gamma_{11} & \Gamma_{12} & \Gamma_{13} \\ \Gamma_{21} & \Gamma_{22} & \Gamma_{23} \\ \Gamma_{31} & \Gamma_{32} & \Gamma_{33} \end{bmatrix} \begin{bmatrix} \int v_1 \, dt \\ \int v_2 \, dt \\ \int v_3 \, dt \end{bmatrix}
\tag{6.8}
$$

The sums of the entries in each row

$$\Gamma_{10} = \Gamma_{11} + \Gamma_{12} + \Gamma_{13}$$

$$\Gamma_{20} = \Gamma_{21} + \Gamma_{22} + \Gamma_{23}$$

$$\Gamma_{30} = \Gamma_{11} + \Gamma_{32} + \Gamma_{33}$$

correspond to inverse shunt inductances in an equivalent circuit of the transformer. If we use the normal transformer equivalent circuit as a T-circuit, the series branches of the T are the leakage inductances and the shunt element is the magnetizing inductance. The Γ's are the inverse admittances in an equivalent Π network. The transfer inverse inductances Γ_{ik} are dominated by the leakage inductances, while the driving point inverse Γ_{io} are dominated by the magnetizing inductances. The algorithm consists of computing the shunt inverse inductances Γ_{10}, Γ_{20}, Γ_{30} from the measured voltages and currents and the known leakage inductances and testing the computed values against a threshold. For example,

$$
\Gamma_{10} = \Gamma_{12} + \Gamma_{13} + \left\{ I_1 - \left(\Gamma_{12} \int v_1 \, dt + \Gamma_{13} \int v_2 \, dt \right) \right\} \bigg/ \int v_1 \, dt
$$

where trapezoidal integration is used to evaluate the integrals. The computed values of the inverse shunt inductances can then be used as the flux is used, to restrain the percentage differential relay without having to compute the harmonic content of the currents. It should be noted that the driving point inverse inductances Γ_{io} are multiple manifestations of the single magnetizing inductance, and hence computation of all of the Γ_{io} represents unnecessary repeat calculations of the inverse magnetizing inductance. In addition the procedure requires the measurement of all of the voltages.

6.2.4 *A restraint function based on the gap in inrush current*

A somewhat unusual restraint technique for detecting inrush phenomena in power transformers has been proposed in reference.[15] It is noted that under certain types of faults there may be significant amounts of second harmonic in the CT secondary currents, which may lead to very long trip times for such faults. The solution proposed by the authors is to identify the condition of magnetizing inrush by detecting the presence of low current (a 'gap') when the transformer core is not in saturation. It is noted by the authors that during inrush conditions the core is not saturated for at least one quarter of a cycle. This gap in the differential current waveform is detected, and when present it is used to restrain the relay from operating. The authors state that this restraint principle avoids the delay in tripping caused by the presence of second harmonic in the differential current for certain fault types. It is stated that field experience with this relaying principle has been successful.

It is, of course, possible to adapt this restraint principle in computer relays. The detection of a gap of quarter cycle in the current waveform will require that a fairly high sampling rate be used so that a sufficient number of samples will correspond to the current gap. If anti-aliasing filters are employed to process the current waveform, their effect on the ability to detect the current gap would also have to be considered. There is no record of a computer relay implementation using this relaying principle.

6.3 Generator protection

Protection of generators with computers has not received as much attention as the protection of lines or transformers. To begin with, the number of generator protection systems on a power system is small, since there are few generators on a power system. Consequently, there is not as big an economic incentive to replace existing relaying with computer relays in this case. Secondly, the burden of protecting a generating station often falls on a plant control system which must deal with the boiler, turbine, generator, and exciter system as a whole; and often is already computer based. Many protection systems in a generating plant are mechanical in nature. The electrical side of the protection includes the stator and rotor winding protection, and several other protection and alarm functions which are closer to being control functions. These have been discussed in Chapter 2, and will be briefly mentioned again later in this section.

6.3.1 Differential protection of stator windings

Early work in this area was constrained by the capability of the available microcomputers.[16] In the present stage of computer development, percentage differential protection of stator windings is a relatively easy task. Assuming that

Figure 6.6   Currents at the two ends of a generator winding

simultaneous samples of currents at the two ends of a stator winding are obtained (see Figure 6.6), a sample by sample percentage differential relation could be used to detect the presence of a fault:

$$i_d(k) = i_1(k) + i_2(k)$$
$$i_r(k) = i_1(k) - i_2(k)$$
$$i_d(k) \leq Ki_r(k) \quad \text{do not trip}$$
$$i_d(k) > Ki_r(k) \quad \text{trip}$$

(6.9)

where K is the slope of the percentage differential characteristic. If the estimation of differential and restraint currents is made on a per-sample basis as in Equation (6.9), it would be necessary to take a vote among several samples to confirm that a fault has indeed occurred, and that the differential current sample is not created by an anomalous data sample for one of the currents.

A somewhat more secure decision is obtained if phasors estimated from $i_1(k)$ and $i_2(k)$ are used in this comparison. Either a mimic circuit must be used to eliminate the influence of the DC offset in the current, or a full cycle DFT technique must be used to calculate the phasors. The DC offset in the fault current of a generator is almost certain to decay very slowly, hence a one cycle DFT would be immune to errors created by DC offset. It is clear that a one cycle DFT makes for a relay which operates in about one cycle, but this may be fast enough in most cases. The percentage differential equation in terms of the phasors is similar to the sample version of Equation (6.9):

$$I_d = |I_1 + I_2|$$
$$I_r = |I_1 - I_2|$$
$$I_d \leq KI_r \quad \text{do not trip}$$
$$I_d > KI_r \quad \text{trip}$$

(6.10)

Saturation of current transformers – although possible – is not a probable event in the case of generator differential protection, as the current transformers are sized generously to avoid saturation in the first couple of cycles. If a transient monitor (see Section 4.5) is used to inhibit operation when a current transformer may be saturated, the phasor comparison will not be made under conditions of CT saturation, and the worst that could happen is that the relay will be slow to respond for internal faults if the current transformers should saturate. A small amount of saturation could be accommodated with a variable slope differential relay, as described in Section 6.5. If differential relaying using samples is being used, care must be taken to terminate the sample differential check when saturation is detected (for example by the transient monitors).

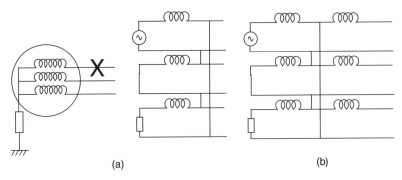

Figure 6.7 A generator with a ground fault. (a) Generator representation and external fault. (b) Internal fault. The interconnection of the sequence diagrams takes place inside the machine boundary in the latter case

6.3.2   Other generator protection functions

Although the generator differential relay is far more sensitive than its counterpart used in transformer protection, a ground fault close to the generator neutral would not be detected if the fault current were less than the sensitivity (slope K) of the differential characteristic. A sensitive ground fault detection scheme for computer relaying of generators has been proposed (although there is no field experience of such a scheme); the relaying scheme is based on the fact that when the stator currents in a synchronous machine are unbalanced, the field current acquires an induced current of twice the fundamental frequency.[17] If the field current is sampled (which can only be done easily when the field current to the generator is fed through slip rings), its second harmonic component can be calculated and used to indicate the presence of stator current unbalance. It then remains to verify that the unbalance is due to an internal fault, and not to some cause that is external to the generator. Consider the generator shown in Figure 6.7(a). For the external fault shown, the negative sequence current flows into the generator as viewed from its terminals; in other words, the fault – which is the source of the negative sequence voltage – is external to the generator. When the fault is inside the machine as in Figure 6.7(b), the negative sequence current flows out from the generator. The 'in' and 'out' of course imply a directional calculation. In the reference cited, the direction of the negative sequence power is used as the restraining function: if the power flows into the machine, the fault is external, and the presence of second harmonic in the field current is disregarded. If, on the other hand, the negative sequence power flows out of the generator, the presence of a second harmonic current indicates an internal fault. By looking at the circuit diagrams in Figure 6.7, it should be clear that, for an external fault, the negative sequence current will lag the negative sequence voltage by almost 90°, whereas, for an internal fault, the corresponding current will lead the voltage, with a substantial in-phase component due to the system load. It seems reasonable that the use of negative sequence reactive power flow may be a better indicator of the direction of the fault than the active power. In any case, both the restraining quantity (negative sequence real or reactive power) and the tripping quantity (second harmonic current in the field winding) must reach a value above a threshold before they can be taken into account, in order to allow for normal residual unbalances and measurement errors. It should be noted that since this technique depends upon load current, it does not offer any protection when the generator carries no load current.

Negative sequence current in a generator is also useful in designing a protection function to guard against excessive rotor heating during system unbalances. The cause of the unbalance is usually outside the generator: often it is a system phenomenon. The capability of a rotor to tolerate negative sequence current is defined by the generator manufacturer in terms of $\int i_2^2 dt$, and it is a simple matter for a computer based relay to incorporate a thermal model of the rotor, and monitor the calculated rotor temperature rise, to determine a more accurate operating limit for the negative sequence current.

Other generator protection tasks, such as loss of field protection, reverse power protection, inadvertent energization protection, volts per Hz protection, out-of-step protection, are all based upon established principles. Computational techniques described earlier, viz. phasor calculation and frequency calculation, are adequate for all these tasks. As mentioned earlier, these functions are more in the nature of control functions, and often the hardest part is to determine what should be done to cope with an event. How to detect this condition is a much easier task.

It should also be noted that a generator protection system is subject to variable frequency conditions during start-up and shut-down, and performance of all algorithms at radically different frequencies must be established. Phasor calculations are affected by the frequency: a phasor calculated by the recursive DFT formula will rotate in the complex plane at a frequency equal to the difference between the nominal system frequency and the actual frequency of the signal (see Chapter 8). Indeed, if the frequency is substantially different from the nominal frequency, the magnitude and rotational speed of the phasor will be variable even though the sinusoid may be of a fixed magnitude and frequency. If the positive sequence phasor is computed, these fluctuations in magnitude and speed disappear, and phasor computations can once again be used.

6.3.3 Sampling rates locked to system frequency

A better alternative may be to use the speed of the generator as an input to the relay system, and use a phase locked loop to adjust the sampling rate to be a multiple of the actual system frequency.[18,19] The phasor calculations made with DFT techniques are exact, and do not require special handling to correct for the effect of off-nominal frequency. It should be noted that phase locked loops used in determining sampling clock rates require some time to acquire and track the system frequency, and sufficient safeguards must be built in to avoid instability of the frequency tracking system due to step changes in the input waveforms. Computer relays based on such principles have been reported in the references cited above.

6.4   Motor protection

Motors rated from 600 volts to 4800 volts are usually protected by fuses, while those rated at 2400 volts to 13 800 volts are protected by relays.[18] Time overcurrent relays used for motor protection are no different from those used for feeder protection. In providing protection for the larger motors with a computer, the currents and voltages must be sampled, and phasors and symmetrical components calculated. Time overcurrent relaying can then be done by taking the magnitude of the current phasors. Another type of protection used for motors consists of protection against an unbalanced source of supply. This can be provided by taking the magnitude of the negative sequence current (or voltage) as a measure of the source unbalance, and if it is found to exceed a pickup setting, the motor is tripped.

An interesting aspect of induction motor protection is to make the protection respond to the temperature rise in the windings, rather than to the stator current as would be provided by a time overcurrent relay.[19,20] The computer offers the ability to create a thermal model of the stator and the rotor, which provides a winding temperature estimate upon which the determination of a safe operating condition can be based.

Time overcurrent relays must be set to trip for a locked rotor condition, yet must tolerate the starting current for the time it takes for the motor to reach its operating speed. If a motor starts with a high inertia load, then the overcurrent relay may trip because of the longer time taken by the motor to reach its running speed (and normal current). Similarly, it is difficult to set the overload relays, because the capacity of the motor to tolerate overload is dependent upon the temperature of the windings before the overload is applied. It is possible to calculate the power loss in the stator and rotor windings from the terminal voltages and currents, and the machine constants. This power loss is used in a thermal model which takes into account the thermal capacity of the machine, its heat loss, and the ambient temperature. In one implementation of such a model,[20] a temperature sensor input is also provided to make the thermal model more accurate, or even unnecessary. It should be noted that these features are available in many conventional protection systems.

As with generator protection, motor protection is quite intimately connected with motor control. Microprocessor based integrated protection and control packages are becoming common in the industry.

6.5   Digital bus protection

Computer relaying of busbars attracted early attention,[15,21] and then the interest flagged until recent times, when busbar protection became a part of an integrated protection and control system for the entire substation.[22-24] Bus protection in an integrated system seems particularly appropriate, as all the inputs needed for bus

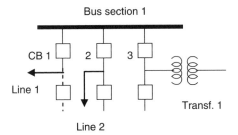

Figure 6.8 A portion of a breaker-and-half substation. The bus protection computer for bus section 1 can obtain the required currents from other protection computers

protection (currents in all circuit breakers and switches connected to the bus) are usually available within all other protection systems in the substation. Consider bus section 1 shown in Figure 6.8, where the protection computers for lines 1 and 2, and transformer 1, use the currents in circuit breakers CB1, CB2, and CB3. These current samples could be shared by their respective protection systems with a bus protection computer through computer-to-computer links. The questions of reliability and redundancy of equipment must be addressed separately. Some consideration of these latter issues will be found in Chapter 9.

As mentioned in Section 2.5, the design of a bus protection system is dominated by consideration of the current transformer performance. First of all, bus differential relaying requires that all current transformers have identical turns ratios, an objective not easy to meet under all circumstances. Any mismatch between CT ratios must be compensated by auxiliary current transformers which add their own errors to the bus protection system CT mismatch error. In a computer relay, auxiliary current transformers are not needed as any main CT ratio mismatch could be corrected in software. However, a much more serious concern is the saturation of a CT for an external fault (see Section 2.5). The very elegant solution offered by a high impedance bus differential relay cannot be used in a computer based protection system, as each current is acquired individually: no analog sum of the feeder currents is formed. (This could be done, but would defeat a significant cost benefit that results from sharing the input information at the computer level.) A new approach to the problem of bus protection in an integrated system must be found.

Disregarding the problem of CT saturation for the moment, it is clear that a percentage differential relay, either based upon sample-by-sample comparison of all the currents, or upon current phasors, can be used. The current phasors – having significant filtering – provide a sensitive and accurate relaying scheme. The latter is also a slower scheme: a reliable phasor estimate must be formed over a reasonable data window before differential and restraining current phasors can be determined. A combination of a phasor based and sample based percentage differential relaying scheme has been included as a component of the bus protection package in one commercially available system.[23] The phasor based scheme could be used as long as there is no significant CT saturation: either early on during a fault before saturation sets in, or much later after the CTs come out of saturation. The transient monitor

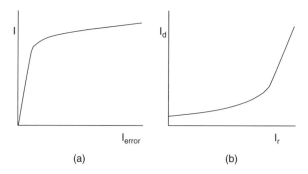

Figure 6.9 Current transformer magnetizing characteristic. (a) Steady state secondary current I produces a secondary voltage V, and the magnetizing current I_{error}. (b) A nonlinear percentage differential characteristic

function is a convenient indicator of the state of the CT. A quarter cycle phasor calculation coupled with a quarter cycle transient monitor would provide a suitable computer based bus differential relay.

An innovation in percentage differential relaying is worth mentioning at this point.[24] In general, a percentage differential relay operates at a constant slope, under the premise that the CT error (when no transient saturation is considered) is proportional to the primary current. On the other hand, the CT saturation characteristic clearly shows – see Figure 6.9(a) – that as the CT voltage increases due to an increase in the secondary current, the errors increase disproportionately. It is clear that such an error performance calls for a percentage differential characteristic with a progressively increasing slope, as shown in Figure 6.9(b) This variable slope characteristic would be desirable for both the sample based and the phasor based percentage differential algorithm. It is provided in a crude form in many conventional bus differential relays, and should be considered for all computer based differential relays which use magnetic core current transformers.

A common concern with all differential protection is that the CT saturation may set in a few milliseconds after the inception of a fault. Rather than devising extremely short-window DFT algorithms, the sample based percentage differential characteristic could be used to provide protection within the first few milliseconds. Normally one would prefer to have at least three samples to make a secure decision in order to eliminate the possibility of being affected by a single bad data point. For external faults, where the percentage differential relay produces restraints, it would be secure to base such a decision on two or even on one sample: as it is almost impossible that in case of an internal fault, a bad data point could be of exactly the right magnitude and polarity to produce a no trip decision. For internal faults, the decision to trip must be confirmed by a concurrence of three or more data points. These ideas are illustrated in the flow chart of Figure 6.10.

The sample based differential relay must be disabled as soon as one of the CTs saturates. As mentioned earlier, the saturation onset could be detected by the transient

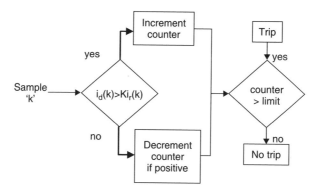

Figure 6.10 A flow chart for a sample based percentage differential relay. To trip, the counter must exceed a certain value. The logic is disabled when saturation is detected

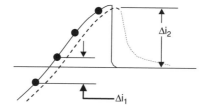

Figure 6.11 Logic to determine CT saturation. The anti-aliasing filter influences the level setting needed by the logic

monitor function. Alternatively, one could check for the change in the secondary current in one sample time (see Figure 6.11). Δi_1 is the change in current during the fault while the CT is unsaturated. This is always less than some value determined by the maximum bus fault current. If the largest expected symmetrical bus fault current from a feeder is I_f, then

$$|\Delta i_1| < \sqrt{2}I_f \times 2\sin\frac{\omega\Delta t}{2} \equiv \Delta i_{max} \qquad (6.11)$$

where ΔT is the sampling interval. When the CT saturation sets in, the change in current is far more abrupt, and a Δi test at each sample could be used to detect the onset of a transient if

$$|\Delta i_k| < M\Delta i_{max'} \text{ no saturation}$$

$$\leq M\Delta i_{max'} \text{ saturation onset} \qquad (6.12)$$

where M is an appropriate margin factor to be set somewhat greater than 1.0. Upon detection of saturation onset, the sample based differential algorithm would be disabled. The anti-aliasing filters will cushion the collapse of the secondary current when the CT goes into saturation, as shown by the dotted line in Figure 6.11.

This effect must be considered to check whether the test defined by inequalities in Equation (6.12) can be useful. Clearly this softening effect is more pronounced when low sampling rates are used: they require lower cut-off frequencies in the anti-aliasing filter. One solution to this problem may be to use a higher sampling rate – indeed, this should always be the goal for obtaining an improvement in the speed of response of all relays. Higher sampling rates will also provide many more samples before the onset of saturation in the bus relaying application. Alternatively, one could obtain samples of the unfiltered currents (i.e. without the anti-aliasing filters) to determine the onset of saturation. Yet another possibility would be to use analog circuits which would provide a trigger indicating the start of CT saturation. The analog mimic circuit discussed in Chapter 5, being essentially a differentiating circuit, would provide such a trigger. The digital mimic circuit would also provide the same function. Return of the CT to an unsaturated state should be confirmed by the purity of its output waveform. Once again, this could be done through a transient monitor function.

A completely different approach to bus protection could be realized through the application of the traditional phase comparison or directional comparison relaying. The direction of the current with respect to a polarizing voltage signal, or an appropriate residual current could be used to determine if all feeder currents from a bus flow into the bus, indicating the presence of a fault. Similarly, one could use the start of a half cycle of current waveform to initiate a phase comparison among all the currents connected to a bus. Since only the start of a current half cycle is used to determine the phase relationship, the fact that one of the CTs may go into saturation is of no immediate concern. Although such schemes permit successful relaying in the presence of early CT saturation, the protection provided is not as sensitive as proper differential relaying – which uses magnitude as well as phase information. Finally, it should be recalled that a great deal of supervision of bus arrangements and switching schemes is often a part of bus protection systems. These functions are far more simple in a computer based bus protection scheme.

6.6 Summary

In this chapter we have examined algorithms for the protection of devices where the percentage differential characteristic is appropriate. In power transformer protection the fundamental issue is providing a restraining function which is capable of recognizing magnetizing inrush and over-excitation conditions. Algorithms for harmonic restraint of the transformer differential protection compute harmonics of the current waveforms. The DFT calculations of the second and fifth harmonic of the restraint current for each phase can be made practical by proper choice of sampling rate.

Transformer algorithms which use bus voltage measurements are possible in an integrated substation protection system. Direct use of the bus voltage as a restraint (restraining if the voltage is high) is possible and produces a potentially faster algorithm than the harmonic restraint algorithm. The bus voltage measurements can also be used to determine the slope of the flux-current characteristic of the transformer. The resulting algorithm is considerably simpler than the Fourier calculations involved in computing harmonics or current and voltage phasors. Finally, although machine and bus protection are not yet as developed as transmission line relaying, we have discussed some of the existing relays that have been designed or are being investigated.

Problems

6.1 The cross-correlation with sinusoids[3] techniques computes the n harmonic in the following manner

$$S_n = \frac{1}{N} \sum_{k=1}^{N} i_k \sin(2\pi(k-1)n/N)$$

$$C_n = \frac{1}{N} \sum_{k=1}^{N} i_k \cos(2\pi(k-1)n/N)$$

$$I_n = 2\sqrt{S_n^2 + C_n^2}$$

Compare the result with the non-recursive DFT calculation.

6.2 The finite impulse response algorithm[4] forms the fundamental and second harmonic from

$$S_1 = i_1 + i_2 + i_3 + i_4 - i_5 - i_6 - i_7 - i_8$$
$$C_1 = i_1 + i_2 - i_3 - i_4 - i_5 - i_6 + i_7 + i_8$$
$$S_2 = i_1 + i_2 - i_3 - i_4 + i_5 + i_6 - i_7 - i_8$$
$$C_2 = i_1 - i_2 - i_3 + i_4 + i_5 - i_6 - i_7 + i_8$$
$$I_n = (\pi/16)\sqrt{S_n^2 + C_n^2}$$

Compare the results with the Walsh calculations. See Figures 3.4 and 4.12.

6.3 The so called rectangular Fourier transform is defined as

$$\overline{S}_n = \sum_{k=0}^{N-1} i_k \, \mathrm{sign}(\sin(2\pi kn/N))$$

$$\overline{C}_n = \sum_{k=0}^{N-1} i_k \, \mathrm{sign}(\cos(2\pi kn/N))$$

If the fundamental and harmonic terms are defined as[5]

$$S_1 = \overline{S}_1 - (1/3)\overline{S}_3 - (1/5)\overline{S}_5$$
$$C_1 = \overline{C}_1 + (1/3)\overline{C}_3 - (1/5)\overline{C}_5$$
$$S_n = \overline{S}_n, C_n = \overline{C}_n \text{ for } n = 2 \text{ \& } 5$$

compare with the Walsh expansion.

6.4 Find the two columns for Table 6.1 that correspond to the third harmonic.

6.5 Reproduce the development of Equations (6.4) and (6.5) with a winding resistance r.

6.6 Assume a nonlinear magnetization curve for a current transformer. Show that a percentage differential relay with fixed slope characteristic will either mis-operate, or be insensitive, depending upon whether the percentage slope is adjusted to correspond to the unsaturated or saturated portion of the magnetization curve.

6.7 Assume that a generator is feeding a one per unit load. For an internal ground fault half-way between the phase *a* terminal and neutral, determine the amount of negative sequence power flow at the machine terminals. Assume that the machine load is entirely passive. What is the negative sequence power flow for an external fault? What is the reactive power flow for each of these cases? Use appropriate constants for the generator.

6.8 Determine the current settings specified in Equations (6.11) and (6.12) for a current transformer with a turns ratio of 1000:5, and a maximum primary fault current of 50 000 amperes (symmetrical). Will these settings work if the anti-aliasing filter shown in Figure 1.8(a) is used in the current input channel?

References

[1] Rockefeller, G.D. (1969) Fault protection with digital computer, IEEE Trans. on PAS, vol. 88, no. 4, pp. 438–461.
[2] Sykes, J.A. and Morrison, I.F. (1972) A proposed method of harmonic restraint differential protection of transformers by digital computer, IEEE Trans. on PAS, vol. 91, no. 3, pp. 1266–1276.
[3] Malik, O.P., Dash, P.K., Hope, G.S. (1976) Digital protection of a power transformer, IEEE Publication 76CH1075-1 PWR Paper A76-191-7, IEEE PES Winter Meeting, January 1976, pp. 1–7.
[4] Larson, R.R., Flechsig, A.J., Schweitzer, E.O. (1977) An efficient inrush current detection algorithm for digital relay protection of transformers, Paper A77-510-1, IEEE PES Summer Meeting, 1977.

[5] Rahman M.A. and Dash, P.K. Fast algorithm for digital protection of power transformers (1982) IEE Proceedings – C Generation, Transmission and Distribution, vol. 129, Part C, no. 2, pp. 79–85.

[6] Degens, A.J. (1982) Microprocessor-implemented digital filters for inrush detection, *Electrical Power and Energy Systems*, vol. 4, no. 3, pp. 196–205.

[7] Thorp J.S. and Phadke, A.G. (1982) A microprocessor-based three-phase transformer differential relay, IEEE Trans. on PAS, vol. 102, no. 2, pp. 426–432.

[8] Habib M. and Marin, M.A. (1987) A comparative analysis of digital relaying algorithms for the differential protection of three-phase transformers, PICA, May 1987, Montreal, Canada.

[9] Einvall C.H. and Linders, J.R. (1975) A three-phase differential relay for transformer protection, IEEE Trans. on PAS, vol. 94, no. 6, pp. 1971–1980.

[10] Harder, E.L. and Marter, W.E. (1948) Principles and practices of relaying in the United States, *AIEE Transactions*, vol. 67, Part II, pp. 1005–1022.

[11] Thorp J.S. and Phadke, A.G. (1982) A microprocessor based voltage-restrained three-phase transformer differential relay, Proceedings of the South Eastern Symposium on Systems Theory, pp. 312–316.

[12] Sykes, J.A. (1972) A new technique for high speed transformer fault protection suitable for digital computer implementation, IEEE PES Summer Meeting, 1972.

[13] Phadke, A.G. and Thorp, J.S. (1983) A new computer based, flux restrained, current differential relay for power transformer protection, IEEE Trans. on PAS, vol. 102, no. 11, pp. 3624–3629.

[14] Inagaki, K., Higaki, M., Matsui, Y. *et al.* (1987) Digital protection method for power transformers based on an equivalent circuit composed of inverse inductance, IEEE PES Summer Meeting, San Francisco, July 1987.

[15] Giuliante, T. and Clough, G. (1991) Advances in the design of differential protection for power transformers, Georgia Tech Protective Relaying Conference, Atlanta, GA, pp. 1–12.

[16] Sachdev, M.S. *Computer Relaying*, IEEE Tutorial, Special Publication no. 79 EH0148-7- PWR, 1979, Chapter 6.

[17] Dash, P.K., Malik O.P. and Hope, G.S. (1977) Fast generator protection against internal asymmetrical faults, IEEE Trans. on PAS, vol. PAS-96, no. 5, pp. 1498–1506.

[18] Benmouyal, G. An adaptive sampling interval generator for digital relaying (1989) *IEEE Power Engineering Review*, vol. 9, issue 7, pp. 45–46.

[19] Adamiak, M.G., Dhruba, P.D., Gardell, J., *et al.* (1993) Performance assessment of a new digital subsystem for generator protection, Twentieth Annual Western Protective Relay Conference, October 19–21, 1993, Spokane, Washington.

[20] Concordia, C. (1951) *Synchronous Machines, Theory and Performance*, General Electric Company.

[21] *Applied Protective Relaying* (1976) Chapter 7, Westinghouse Electric Corporation.

[22] Zocholl, S.E., Schweitzer III, E.O., and Aliaga-Zegarra, (1984) A. Thermal protection of induction motors enhanced by interactive electrical and thermal methods, Trans. of IEEE on PAS, vol. PAS-103, no. 7, pp. 1749–1755.

[23] Zocholl, S.E. Determining relay settings for induction motor stator and rotor protection using thermal models (1987) Conference on Computer Relaying, Blacksburg, Virginia.

[24] Cory B.J. and Moont, J.F. Application of digital computers to busbar protection (1970) IEE Conference on the Application of Computers to Power System Protection and Metering, Bournemouth, England, pp. 201–209.

[25] *Microprocessor Relays and Protection Systems*, IEEE Tutorial Course, Special Publication 88EH0269-1-PWR, Winter Power Meeting of the PES, 1988.

[26] Udren E.A. and Sackin, M. Relaying features of an integrated microprocessor-based substation control and protection system (1980) IEE Conference Publication 185, *Developments in Power System Protection*, London, pp. 88–92.

[27] Udren, E.A. (1985) An integrated, microprocessor based system for relaying and control of substations: Design features and testing program, 12th Annual Western Protective Relaying Conference, Spokane, Washington.

7

Hardware organization in integrated systems

7.1 The nature of hardware issues

In Chapter 1, we discussed several hardware related questions – such as the computer hierarchy in the substation, subsystems of a computer relay, and the analog to digital converters. In this chapter, we will explore in greater detail some of the characteristics, environment, and maintenance issues of computer relays which are crucial to the success of field installations of these devices.

The computing power of microprocessors has been increasing steadily and dramatically over the past several years. It is therefore not particularly useful to describe and lock-in on the computer hardware capabilities of devices used in relaying at present. On the other hand, one needs to be aware of some considerations which stem from functional needs of relaying, and from relaying application considerations it is possible to specify the hardware needs. One could view such a discussion as pointing to the minimum acceptable capability required for satisfactory relaying programs – and it follows that, as computers get better, they will meet the relay task requirements with even greater margins of computational capacity. We will make an attempt to arrive at a description of the minimum hardware capability needed for computer relays.

The three-level hierarchy of computers within the substation was shown in Figure 1.12. Level I computers are the protection computers, and usually are placed inside the substation control house. It is likely that in the future the relays may reside outdoors next to the power apparatus. Should this be the case, the environmental conditions in which the relays are placed would be far more demanding than they are at present. Level II computers are central to the substation, and are therefore placed in the control house also. It is extremely important that

Computer Relaying for Power Systems 2e   by A. G. Phadke and J. S. Thorp
© 2009 John Wiley & Sons, Ltd

the physical environment inside the control house be carefully defined, so that both the designers of the relay equipment on the one hand, and designers of substations and control houses on the other, produce installations in which the capability of the relays is matched to the conditions that exist inside the control house. We will describe the station environment and quantitative specifications that have been used for computer relays.

A somewhat different issue is that of hardware reliability through redundancy. We will examine the traditional practice of achieving dependability through duplication of hardware. However, other avenues for achieving greater dependability (and perhaps security) are available with computer relays. We will consider these issues later in this chapter.

Finally, we will discuss the question of servicing and maintaining the computer based protection system, and training of station personnel in this new field.

7.2 Computers for relaying

We will now discuss considerations that affect and set limits for acceptable performance for computers to be used for relaying. It has been recognized for some time that impedance relaying remains an important technique for power system protection, and furthermore it encompasses the most computationally demanding algorithms. We will therefore use distance relaying considerations in order to arrive at computational needs.

We begin with the basic feature of a distance relay which affects all aspects of computer relays – the speed of response. It has been shown that the speed of a distance relay is limited by the transient phenomena that accompany a fault, and that considering the entire fault clearing process, it is not significantly beneficial at present to achieve relay speeds faster than a quarter of the period of the fundamental power frequency.[1] Thus the fastest desirable speed of a relay response may be taken to be between four and five milliseconds (depending upon whether the power system is operated at 60 Hz or 50 Hz). To make a secure decision for relaying over this period, more than two data samples are needed. Thus we reach the conclusion that the analog input sampling frequency should be at least 12 times the fundamental frequency of the analog input signals with a sampling interval of the order of 1.4 milliseconds. Since the relaying algorithms must execute between samples, it is not advantageous to increase the sampling rate to double or triple (i.e. to 1440 or 2160 Hz) this rate, as the algorithm execution time would then be reduced to 0.7 or 0.5 milliseconds respectively. As most relaying algorithms seem to be accommodated by a couple of thousand machine language instructions, a sampling period of 0.5 millisecond would call for an *average* instruction time of the order of 250 nanoseconds. In fact such an algorithm would not leave any margin of safety, and a more realistic average instruction time needed for a sampling rate of 2160 Hz would be 100 nanoseconds.

One could thus arrive at the conclusion that, with computers having an average instruction time of 100 to 300 nanoseconds, satisfactory relaying algorithms could be built around sampling frequencies of between 36 times and 12 times the power system frequency. As computer capabilities increase in the coming years, sampling rate used in relaying tasks is also likely to increase.

Perhaps we should recall that, in an integrated computer system, the sampled data may also be used for oscillography. In such cases, it may be desirable to sample at much higher rates to reproduce the higher frequency transients. If higher sampling rates are used in the data acquisition units (see Figure 1.11), the sampled data must be converted to lower frequency samples before relaying algorithms are invoked. The data reduction must also simulate correct anti-aliasing filters that would go with the lower sampling rate.

Having discussed the desirable instruction execution time for the computer and the corresponding sampling rate, we will next consider the computer word length. It is clear that 16 bit microcomputers are commonplace now, and 32 bit computers are becoming available. The 16 bit computation goes well with the 12 bit A/D converters currently in use. If we consider a typical computation involved in a relaying algorithm (for example, the phasor calculations as in Chapter 3), we construct sums of the type

$$X = \sum_{k=1}^{N} a_k x_k$$

The numbers a_k being sines or cosines of sampling angles are at most 1.0; thus the accumulated sum $X \le N x_k$. In the worst case, a set of 12 samples (for a complete cycle of data) with x_k obtained from a 12 bit A/D converter would produce X no greater than a 16 bit number. Of course, in reality, the Fourier sum calls for a normalizing factor of $\sqrt{2}/N$, but this is generally omitted in the interest of economy of computation.

If the A/D converter is a 16 bit converter, at least 20 bits will be needed to contain the phasor, or a scaling procedure will be needed. However, scaling down would throw away some of the precision delivered by the 16 bit converter. Consequently, a 16 bit A/D converter would work well with a 32 bit computer to produce a precision in computation commensurate with the A/D system.

There is some evidence that a 32 bit computer and a 16 bit A/D converter would be very desirable in order to meet the needs of a distance relay which must cope with high currents for near faults, yet for some of the backup functions (or fault detection functions) be responsive to very small fault currents. This is the issue of dynamic range of the currents (and, to a lesser extent, of voltages) for which the relay is responsible. As explained in Section 1.5, the 12 bit A/D converter resolves 1 part in 2048. If the largest possible current is scaled to produce a digital value of 2048 when it is fully offset, the corresponding current without a DC offset would produce a digital value of about half, or 1024. If we may set an acceptable error due to quantization of no more than 1%, then the smallest current that could be measured with this accuracy would have a peak of about 50, since the quantization error is equal to half the least significant bit – or 1/2. Thus, the 12 bit A/D converter may respond to a dynamic current range of (50:1024) or about (1:20) and still produce accurate digital values of current samples to within 1% of their peak value. This dynamic range is not quite sufficient in many situations. In several relay locations, it is desirable that the relay measure currents with dynamic ranges of up to 1:200. This requirement can be met with an A/D converter with 14 bit resolution; and certainly a 16 bit converter would produce adequate accuracy over a dynamic range of 1:800. With such a system, even light load currents as well as maximum fault currents could be read with high accuracy. It should be noted that 16 bit delta-sigma A/D converters are becoming quite common, and are in use in some of the currently manufactured relays.

Computer instruction execution time, word length, and A/D converter resolution are the main features which influence the quality of measurement performed by a relay. Other features which make for good overall performance are immunity to interference, low power consumption (two features that are at odds with each other), adequate peripheral equipment, and assured supply of spare parts. Anticipating the future availability of high speed direct digital transducer (CT and VT) outputs, the computer should also be able to acquire these samples without significant software overhead.

7.3   The substation environment

Physical conditions within an electric utility transmission substation are among the most severe that can be imagined. In the outdoor station yard, the temperatures can become very high: 120°F (49°C) in very hot regions of the world during summer, and very cold: 60°F (−51°C) in cold climates during the winter. At a given site, the annual variation of the temperature may be 150°F (65°C), and within one day the temperatures may change by 70°F. The control house is usually a covered building with fans and heaters, and is sometimes air-conditioned. Although some substations may be manned office type buildings, most of the time they are unmanned enclosures made of corrugated steel sheets.

In addition to the extremes of temperature and humidity that the substation equipment must withstand, there is also a hostile electromagnetic environment. When switching operations take place in the substation yard – especially with disconnect switches arcing in air – high levels of electromagnetic fields are set up in the yard and control house. Faults occurring within the substation or near it also cause ground currents and ground potential rise which may influence all protection and control equipment within the station. In addition, various high voltage apparatus may have corona discharges of varying severity depending upon the weather. Furthermore, relay operation within the control house may generate transient fields which may affect other relays and control equipment. And finally, the relays and other protection and control equipment may be affected by fields produced by hand-held walkie-talkie type radio communication equipment.

The control house – where most protection equipment is located – generally provides considerable shielding from radiated interference originating in the substation switchyard. Careful wiring practices and equipment shielding techniques must be employed in designing relaying equipment in order to ensure that the equipment will perform satisfactorily under all reasonable service conditions. The substation designer and the protection designer use industry standards to arrive at environmental specifications which are not exceeded inside the control house, and are exceeded by the immunity built into the relays. Each country has an appropriate standard to follow. Basically these standards specify a level for conducted Electromagnetic Interference (EMI), and some guides for temperature, humidity and radiated interference. More recently, attention is being given to the Electromagnetic Pulse (EMP) fields produced by nuclear warheads, but no standards exist as yet for the EMP fields. We will discuss a few of the relevant standards and guides in the next section.

7.4   Industry environmental standards

An industry standard provides operating temperature and humidity specifications for supervisory control, data acquisition and substation automatic control equipment.[2] Although later revisions of this standard have indicated that these specifications may not apply to protective relays, we could use the standard as a typical environmental specification document. For buildings without air-conditioning, the applicable temperature range is 0°C to 55°C with an allowable rate of temperature change of 20°C per hour. The humidity range is specified to be 10–95% without condensation. For outdoor equipment (for example in the substation yard), the applicable temperature range is −25°C to 60°C. The 1974 ANSI/IEEE standard C37.90a for relaying equipment specifies the permissible ambient air temperature close to the relays to be −20°C to 55°C.[3] A computer relaying system built for service in the field has used a temperature specification of 0°C to 55°C.[4] Standards for seismic shock withstand are defined by another standard.[5]

The control house within a substation provides substantial shielding from radiated EMI. The wiring from the switchyard to the control house penetrates this shield, and consequently the EMI induced in this wiring and conducted to the protection equipment remains a major source of concern. The ANSI/IEEE standard C37.90a of 1974 and its revision[6] provide the standard for Surge Withstand Capability (SWC) to be built into protective equipment.

Consider the relay system shown schematically in Figure 7.1. Its connections to the outside world are through the four groups of circuits shown: analog inputs (voltages and currents), digital inputs (through which external contact or switch status is communicated to the relay), digital outputs, and power supply. EMI can reach the relay through any of these wires. Consequently, a specification is provided

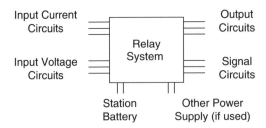

Figure 7.1   Relay input-output definition for Surge Withstand Capability test

for EMI level at the terminals where these wires enter the relay system. The EMI may be transverse (differential mode), or longitudinal (common mode). These two terms refer to interference voltages between any pair of wires (transverse mode) and all wires and ground (common mode). The specifications are for a group of wires at a time while the relay is operating normally. Thus, while the relay inputs and outputs are connected normally, one group of wires (for example the current input leads) is subjected to the SWC test. A typical set-up for these tests is as shown in Figure 7.2. The surges are coupled through capacitors, and they are restricted to the relay under test by blocking inductors in series with each wire.

The standard SWC test consists of two parts: (1) the oscillatory SWC test, and (2) the fast transient test. The oscillatory test attempts to duplicate the EMI induced in the wiring due to the switching operations or faults within the substation. It is defined as an oscillatory wave at the terminals of the SWC generator in the frequency range of 1.0 MHz to 1.5 MHz, with the first peak value of 2.5 to 3.0 kV crest, and the envelope decaying to 50% of the first peak in not less than 6 microseconds. The source impedance of the test source (accounting for the energy in the wave) should be in the range of 150–200 Ohms. Further, the test wave is to be applied at a rate of at least 50 applications per second for a minimum duration of 2.0 seconds. These specifications are illustrated in Figure 7.3. The test frequency, energy content,

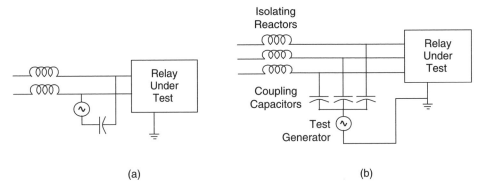

(a)   (b)

Figure 7.2   Transverse and longitudinal SWC testing of relay current input circuits. (a) Transverse mode. (b) Longitudinal mode

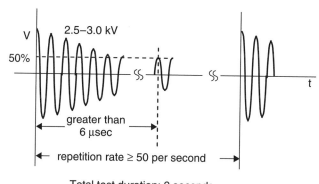

Figure 7.3   The oscillatory SWC test specification

repetition rate etc. for the oscillatory SWC test have been arrived at after reviewing a number of field observations in HV and EHV substations.[4,6]

The fast transient SWC test has been added because it was found that these types of surges are introduced in the relay wiring whenever low currents in auxiliary relay coils within the control house are interrupted. The contacts used for interrupting the current may restrike repeatedly, causing a compound transient of several restrikes and interruptions to occur until the coil current is finally extinguished. These transients have been proven to be particularly destructive in solid-state relays of early designs. These phenomena are equally significant in determining the capability of the computer relays.

The fast transient SWC test wave is unidirectional. Its rise time – defined to be the time required to reach from 10% of its peak to 90% of its peak value – should be 10 nanoseconds or less. The crest duration should remain above 90% of its peak for at least 50 nanoseconds; and the decay time to 50% of its peak value should be between 100 and 200 nanoseconds. The crest value of the voltage should be between 4 kV and 5 kV. The source impedance of the test wave generator should be 80 Ohms or less. These pulses should be applied at least 50 times per second, for a test duration of at least two seconds. The polarity of the test wave should be both positive and negative. The fast transient waveform specifications are illustrated in Figure 7.4.

Note that both the oscillatory and unidirectional test wave specifications apply to the waveform generator terminals under open circuit conditions. During the test itself, these voltage magnitudes and decay rates will be affected by the relay equipment under test.

It is also intended that the relay input quantities (power frequency voltages and currents) during the SWC test be adjusted to values which put the relay on the verge of operation. It is required that the relay not change its status during the SWC tests. It is of course understood that no permanent damage to the relay should occur as a consequence of the SWC tests.

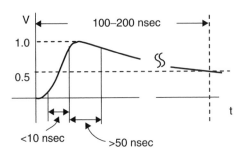

• 50 times per second
• Total duration 2 seconds

Figure 7.4 Fast transient SWC test waveform. This waveform approximates the transients generated by switching an inductive circuit powered by the station battery in the control house and repeated arcing of the interrupting contact

In addition to the ANSI/IEEE test described above, there are other standards from the International Electrotechnical Commission (IEC)[7] and the British Electrical and Allied Manufacturers' Association (BEAMA),[8] and in some cases from large Electric Utility companies. In general, the peak voltages specified by these standards range between 1.5 kV and 6 kV. The test generator source impedance varies between 50 ohms and 500 ohms, and the test frequency ranges between 0.10 and 2 MHz.

In addition to the conducted EMI standard provided by the SWC tests, radiated interference[9] and Electromagnetic Pulse phenomena[10] should also be taken into consideration.

As mentioned earlier, the various specifications and standards for environmental conditions serve two functions: to help design a substation which will produce conditions less severe than those specified by the standard, and to help design equipment which can withstand conditions that exceed the standards. Thus the standards provide a common, well defined compromise position for the manufacturer and user of the protective equipment. It is of course possible to use conditions other than those defined by the standard – provided that the manufacturer and user mutually agree to do so. From time to time, other phenomena and failure modes of protective equipment may point out unforeseen conditions which may lead to a modification of the standards. Therefore the environmental standards should be viewed as dynamic statements about the current understanding of the phenomena involved. In the context of computer relaying, it has not been considered necessary to modify the existing standards.

7.5 Countermeasures against EMI

The design of relays – whether computer based or solid-state – must incorporate SWC filtering on all wires which penetrate the relay enclosure. The SWC filter must reduce the transient surges defined by the SWC standard to acceptable levels within the relay. One such filter was described in Chapter 1. Other filter designs are possible, and will depend upon the capability of the circuit board where the signals must be applied. To that extent, each SWC filter is an integral part of the relay system and is designed to accommodate the specifications of the computer and its peripheral hardware.

The substation design and wiring must also follow a procedure which will limit the EMI signals at relay input terminals to a value specified by the SWC standard.

Perhaps the strongest source of EMI in the substation is a disconnect switch which opens a capacitive current (such as a bus section or coupling capacitor device) and has another capacitance to ground on the source side (again, either a bus section or capacitive voltage transformer). As the current is interrupted (see Figure 7.5), the arcing contacts reignite several times – and each time a high frequency high voltage oscillation is initiated.[11,12] These phenomena produce induced voltages in the control and signal wiring that connects the CT and CVT windings and circuit breaker control wiring to the relay located in the control house. It has been found that shielded wires reduce this coupling significantly.

Both wires of a signal should be carried inside a common shield. The shield should be grounded to the station ground mat at both ends and preferably at as many points along the cable run as practical.[11,13] The signal circuit should be grounded at one point only, generally inside the control house (see Figure 7.6). In addition, special attention should be paid to grounding of the capacitive voltage transformer. Multiple low resistance grounding wires should be used to ground the CVT base and thus lower the surge impedance of the ground connection. The run of the wire trenches should be as far as possible from the sources of EMI. Similar circuits (current

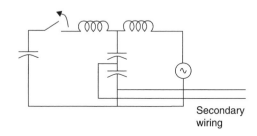

Figure 7.5 Interruption of a capacitive current by a disconnect switch

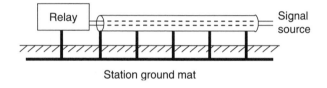

Figure 7.6 Shield grounding on signal and control cables

transformer secondary leads, for example) should be included in the same conduit. Where extreme noise immunity is desired, doubly shielded wiring may be necessary. Shielding should be extended to all wiring that penetrates the control house – low voltage power supply, auxiliary (i.e. non-relaying) control cables should all be shielded and grounded carefully, as these wires can also bring EMI inside the control house where it may do damage.

Surges generated inside the control house – for example by fluorescent lights, switched inductive currents, etc. should be controlled through appropriate means. The inductive current interrupters should be protected by capacitors or by metal oxide varistors of adequate current handling capability.

7.6 Supplementary equipment

For the sake of completeness, we now enumerate other equipment which, although essential to the computer relaying functions, is no different from equipment needed in conventional relaying systems. A computer based protection system must include this supplementary equipment, and some features of the relay architecture may be affected by such equipment.

7.6.1 Power supply

The station service for protection is generally 125 V DC supplied from station battery. The battery is continuously charged by a charger of adequate capacity connected to the AC station power supply. Occasionally other battery voltages such as 48 V or 250 V may be encountered. A computer relay would require DC to DC power converters to bring this battery voltage down to the usual computer requirements of 5 V DC and ± 15 V dc. With sufficient computer equipment within a substation, it may become practical to furnish a battery system which supplies the computers directly through their own chargers.

7.6.2 Auxiliary relays

These are electromechanical or solid-state isolating relays which provide multiple closing and opening contacts which may be used for signaling and tripping duties.

7.6.3 Test switches

At the time of commissioning a relay, and later during its periodic testing and calibration, the relay outputs must be isolated from the breaker trip coils, and at the same time the relay inputs must be disconnected from the current and voltage transformers. This arrangement is shown schematically in Figure 7.7. In conventional relays the test switch is a single device with multiple switching contactors which control all connections to the relay in one operation. In the case of current transformer inputs, the secondary windings of the CT must not be open circuited; hence

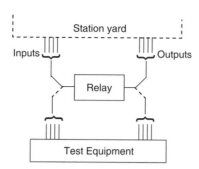

Figure 7.7 Relay test switches isolate the relay from the power system

those contacts must be shorted before the relay is isolated. Similarly, the breaker trip coil and signaling circuits must be left in a secure quiescent state as the relay is removed from service. This is usually accomplished by designing the test switch to provide appropriate make-before-break and break-before-make contacts as needed.

7.6.4 Interface panel

A traditional relay panel consists of control switches, adjustment knobs, visual indicators for various settings, and alarm targets. All present day operator interfaces have a mechanical aspect so that a control action has an associated tactile feeling. The operator gets a sensation of having changed something physically, and he is immediately rewarded by a mechanical display confirming the change he has introduced. Although mechanical linkages may break, dial lights may burn out, and a target may fall erroneously, this type of panel is traditional, and all relay engineers and field personnel are fond of such interface panels.

On the other hand, a computer based relay can accommodate all required operations and responses through an operator's video console and keyboard. Certainly, for such a 'panel', a serial port to which a console can be connected is all that needs to be available on the relay front panel. In the present state-of-the-art some type of an interface panel which emulates the present relay panels is being incorporated in all computer relay designs. In most cases, a console port is also provided, and in principle this console may be available at a remote location through a communication link.

7.7 Redundancy and backup

A redundant protection system provides backup in the event that some misoperation in the relaying system leads to a failure within the relay, or in any of its inputs (CT and CVT secondary circuits). Often a redundant protection uses a different principle of protection – for example, a step distance relay may be the redundant system for a phase comparison system. The redundant system would guard against the failure to trip, making the overall protection more dependable.

In a computer relaying system, one could duplicate the two different relaying principles within the same hardware. One would thus be providing for dependable operation if one of the principles of protection were invalid for certain types of faults. The need for hardware duplication remains, and therefore in a computer relaying system as well one must provide for redundancy. However, as it is a computer based system with communication capability, redundancy could be achieved in many different ways.

One could arrange a single computer relay to serve as duplicate hardware for all the other computer relays in the substation. Such a computer would have all the inputs and outputs brought to it, and would, upon notification of a need to back up a given relay, begin processing the inputs and outputs of that relay under program control. This arrangement is shown in Figure 7.8. It should be clear that the redundant relay in this scheme cannot be as fast as the front-line relay, as it must be alerted to switch its mode of operation to that of the relay in question. The delays involved need be no longer than a quarter-cycle of the fundamental power frequency.

If the input-output systems are computer based, the connections to the relaying processors are over a communication channel. A data highway type of connection is then possible, although from the point of view of common mode failure, such an arrangement may not be satisfactory. Furthermore, data transmission from all the input-output systems situated in the substation to all the relaying processors in the control house would require excessive rates of data transfer. A radial connection from one input-output system to all the designated protection computers would be less demanding of the channel speed (see Figure 7.9). One could provide for the failure of the input-output system computers by duplicating them, although the failure of the current transformer or voltage transformer secondary circuits would disable both input-output systems. One could of course visualize duplication of transducer secondary windings as well.

Another method of backup of the input-output system is possible in a computer based system. One could use the analog data acquired by another computer relay

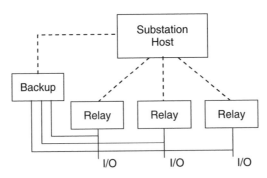

Figure 7.8 A single redundant relay acts as a backup for computer relays in a substation. The backup relay assignment is made by the host computer

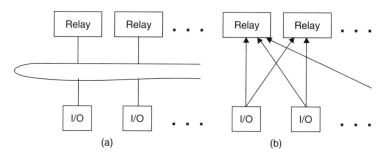

Figure 7.9 Connections between input-output systems and the protection computer. (a) Data highway. (b) Radial connections

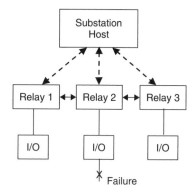

Figure 7.10 Backup for failed analog input system through processor–processor links

to substitute for a failed analog input system. As shown in Figure 7.10, when the input system of relay 2 fails, it could substitute for the missing data by using information from relays 1 and 3. Thus, a missing voltage could be obtained if voltages seen by relays 1 or 3 were electrically the same as those seen by relay 2. Missing currents could be reconstructed by invoking Kirchhoff's current law around a power system bus.

In summary, a computer based protection system offers many alternative ways of providing hardware duplication. Duplication of hardware is still necessary – processors as well as transducers must be duplicated – but for a given duplicated hardware, far more flexibility in back up is achievable in a computer based system. Functional backup, whether remote or local, can be achieved within the same hardware by programming the appropriate functional algorithms.

7.8 Servicing, training and maintenance

We include this section as a reminder that with computer relaying come a host of other issues which must be addressed before these new systems can become acceptable field grade equipment. Certainly the most troublesome is the question of maintainability of this equipment for many years after it is installed. It is well known that traditional (i.e. conventional) electromechanical or solid-state relays have a service life of 10–20 years, and in some cases 40 year old relays are still in service. It does not seem likely that computer based relays can be (or should be) maintained so long. It is a fact that the computer hardware technology advances too rapidly: a five-year-old hardware chip is usually an obsolete one. It is no longer available in the marketplace – neither should it be, because it has been replaced by a superior product. In the face of this rapidly changing technology, the computer based relay is unlikely to be a device with a 40 year life span. A standard computer back-plane, into which a succession of compatible boards may be plugged in the future, looks like a reasonable alternative. In any case, substantial rejuvenation programs for computer based relays and integrated systems look like a concomitant of this technology.

Maintenance of computer relays should be a simpler proposition. With self-diagnostic capability of computer based devices, the defective sub-assemblies should be identifiable. The replacement procedure could be generated on site, with detailed, step-by-step advice to the maintenance engineer. With a common hardware implementation for many – or all – substation computers, the spare part inventory should be small.

The service and field personnel would certainly need training to install, calibrate and repair these systems. There is a very large community of computer hardware installers, repairmen, and diagnosticians. No doubt their expertise will be called upon to train the substation personnel. In any case, although newer training procedures are required, there should be no basic hindrance to this undertaking.

The most interesting aspect of the entire system is the retraining of relay engineers and system planners to adapt to this new technology. The relay engineer could become a designer of relaying functions through the flexible software capability of a computer relay. The manufacturer could furnish the hardware and a software shell, in which the user could place a custom-designed application software. The relays could be altered in the field automatically (as in case of adaptive relays), or through centrally directed commands. The relays – in the event of an erroneous operation – could be queried as to what led to the error. Hidden flaws in relay design logic could thus be uncovered. The sequence of events and oscillography information could be obtained at a central location within minutes of the event. All these possibilities come with computer relaying – and relay engineers of the future must begin to think in terms of these expanded roles for the computer relays.

7.9 Summary

In this chapter, we have dealt with present day hardware suitable for computer relaying. Microcomputers with 16 bit word length and Analog to Digital Converters with 12 bit resolution seem to be adequate for all relaying needs. There is some indication that the next generation of 32 bit computers and 16 bit ADCs could lead to noticeable improvement in measurement functions performed by the computer relays. We have described the available industry environmental standards for temperature, humidity, seismic shock, and electromagnetic interference which a computer relay must meet in the substations. We have also described some of the established methods of combating EMI interference. And, finally, we have given a general review of reliability and backup as influenced by hardware considerations.

References

[1] Thorp, J. S., Phadke, A. G., Horowitz, S. H. and Beehler, J. E. (1979) Limits to impedance relaying, IEEE Trans. on PAS, vol. PAS-98, no. 1, pp. 246–260.

[2] IEEE Standard: Definition, Specification, Analysis of System used for Supervisory Control, Data Acquisition, and Automatic Control, ANSI/IEEE C37.1-1979.

[3] Guide for Surge Withstand Capability (SWC) Tests, ANSI C37.90a-1974, IEEE Standard 472-1974.

[4] Substation Control and Protection Project – System Requirements Specifications, Electric Power Research Institute EL-1813, Interim Report, April 1981.

[5] IEEE Standard Seismic Testing of Relays, ANSI/IEEE C37.98- 1984.

[6] IEEE Standard: Surge Withstand Capability (SWC) Tests for Protective Relays and Relay Systems, P472/D9, C37.90.1-198x. Draft Document of the Power System Relaying Committee, June 8, 1987.

[7] *Single Input Energizing Quantity Measuring Relays with Dependent Specified Time*, International Electrotechnical Commission, IEC Standard, Publication 255-4, first edition, 1976.

[8] *Recommended Transient Voltage Tests Applicable to Transistorized Relays*, The British Electrical and Allied Manufacturers' Association (Inc.), Publication no. 219, November 1966.

[9] Withstand Capability of Relay Systems to Radiated Electromagnetic Interference, ANSI 37.90.2, IEEE Standard P734/D2, Draft Document of the Power System Relaying Committee, February 1981.

[10] The Effects of EMP on Protective Relaying Systems, Report by the EMP Working Group of the Power System Relaying Committee of IEEE, 1986.

[11] Callow, J. A. and Mackley, K. W. Impulsive Overvoltages on Secondary Circuits of 330kV Capacitor Voltage Transformer, Snowy Mountain Hydroelectric Authority, CIGRÉ 1962, Report No. 136.

[12] Dietrich, R. E., Ramberg, H. C. and Barber, J. C. (1970) BPA Experience with EMI measurements and shielding in EHV substations, Proceedings of the American Power Conference, vol. 32, pp. 1054–1061.

[13] Kotheimer, W. C. (1969) Control circuit transients, *Power Engineering*, Part I, vol. 73, pp. 42–45, January 1969; Part II, pp. 54–56, February 1969.

8

System relaying and control

8.1 Introduction

The presence in the substation of microprocessor based devices, which are constantly processing data received from the system, makes it possible to include the protection system as a part of a system-wide computer hierarchy dedicated to monitoring and control. A relay can be regarded as a measuring system. As we have seen in the previous chapters, almost all relaying algorithms begin with samples of the measured system voltages and currents. In many cases phasor voltages and currents are computed before relaying decisions are made. While faults occur extremely infrequently, the phasor calculations are being performed constantly. Because of the redundancy built into the protection system itself and because of the digital equipment's self diagnostic ability, the computer relay becomes a highly reliable measuring device.

The measuring function is a logical complement to the protection function. In fact, the two functions can coexist if proper priorities are maintained. During normal system conditions the measuring function would be active. During fault conditions measuring would be suspended (the measurements are of questionable value during faults) and the measurements used for relaying decisions. The question of measurement accuracy should also be considered. In normal (quasi-steady state) operating conditions the fault induced non-fundamental frequency signals discussed in Chapter 5 are not seen by the relay. In addition, longer window algorithms can be used for the measurement function, further reducing the error in the result (in an inverse time way, as in Figure 5.25). Since measurements are not required at the high rate required for relaying, it is possible to imagine data windows of cycles for the measurement function. The two functions are compatible so long as some trigger, such as the transient monitor, is used to initiate the relaying calculations and suspend the measurement function.

The fact that the relaying processors are connected through the substation host to central computers in a fully integrated system is an additional benefit. Monitoring

Computer Relaying for Power Systems 2e by A. G. Phadke and J. S. Thorp
© 2009 John Wiley & Sons, Ltd

and control functions, both local and central, can have access to measurements from the relaying processors. It is clear the local breaker control – high speed reclosing, automatic reclosing, synchronizing checks, bus reconfiguration, etc. – can be initiated by the host computer on the basis of data received from the protection processors. Additional control functions would profit from the availability of such data from remote substations. For example, VAR control through capacitor switching could be based on voltage information from neighboring buses. Bus reconfiguration could be influenced by the configuration of adjacent buses.

This chapter is concerned with the relaying computer as a measuring device and the use of those measurements in monitoring and control. The next two sections describe a technique for extracting additional value from the phasor measurements produced by the Fourier type algorithms.

8.2 Measurement of frequency and phase

An interesting and useful result can be obtained by examining the SCDFT (Section 5.5) calculations of the positive sequence voltage for a balanced system operating at a frequency different than the nominal power system frequency.[1] If we denote the nominal power system frequency and write the phase voltages as

$$v_a(t) = \text{Re}\{\overline{V}e^{j\omega t}\} = \frac{1}{2}[\overline{V}e^{j\omega t} + \overline{V}^* e^{-j\omega t}]$$

$$v_b(t) = \text{Re}\{\overline{V}\alpha^2 e^{j\omega t}\} = \frac{1}{2}[\overline{V}\alpha^2 e^{j\omega t} + \overline{V}^*\alpha^{*2} e^{-j\omega t}] \tag{8.1}$$

$$v_c(t) = \text{Re}\{\overline{V}\alpha e^{j\omega t}\} = \frac{1}{2}[\overline{V}\alpha e^{j\omega t} + \overline{V}^*\alpha^* e^{-j\omega t}]$$

where $\overline{V}$ is the positive sequence phasor, α is $e^{j2\pi/3}$ and ω is different from ω_o. If we compute the positive sequence voltage with a full cycle window, for example, corresponding to K samples per full cycle of the nominal frequency, ending at sample L, from Equations (5.32) and (5.76)

$$\tilde{V}_1^{(L)} = \frac{2}{K} \sum_{k=L-K+1}^{L} \frac{1}{3}[v_a(k\Delta t) + \alpha v_b(k\Delta t) + \alpha^2 v_c(k\Delta t)]e^{-jk\omega_o\Delta t} \tag{8.2}$$

Using Equation (8.1)

$$\tilde{V}_1^{(L)} = \frac{1}{3K} \sum_{k=L-K+1}^{L} 3\overline{V}e^{j\omega k\Delta t} e^{jk\omega_o\Delta t}$$

$$+ \frac{1}{3K} \sum_{k=L-K+1}^{L} \overline{V}^*(1 + \alpha\alpha^{*2} + \alpha^2\alpha^*)e^{-j\omega k\Delta t} e^{-jk\omega_o\Delta t} \tag{8.3}$$

The last term in Equation (8.3) is zero since $\alpha\alpha^* = 1$, and $(1 + \alpha^* + \alpha) = 0$. If we let $\omega = \omega_0 + \Delta\omega$

$$\tilde{V}_1^{(L)} = \frac{1}{K} \sum_{k=L-K+1}^{L} \overline{V} e^{jk\Delta\omega\Delta t} \tag{8.4}$$

The sum in Equation (8.4) can be evaluated and yields

$$\tilde{V}_1^{(L)} = \overline{V} e^{jL\Delta\omega\Delta t} \left[\frac{\sin\left(\frac{K\Delta\omega\Delta t}{2}\right)}{K\sin\left(\frac{\Delta\omega\Delta t}{2}\right)} e^{-j(K-1)\Delta\omega\Delta t/2} \right] \tag{8.5}$$

The bracketed term in Equation (8.5) represents an error in computing the phasor caused by the power system operating at a frequency other than the nominal frequency. It has both a magnitude and an angle. At a sampling rate of 12 samples per cycle and a nominal power system frequency of 60 Hz, the magnitude of the bracketed term varies from .98869 to 1.0000 as the power system frequency varies from 55 to 65 Hz (a considerable range). The angle of the bracketed term is 2.7 degrees per Hz for the same system. The bracketed term is also unimportant in frequency calculations since it is a constant, i.e. independent of L. For an interesting discussion of the single phase version of these calculations where the second term of Equation (8.3) is nonzero, the reader is referred to the discussion and closure of Reference 1.

The part of Equation (8.5) that depends on the recursion number L is the angle $L\Delta\omega\,\Delta t$. If we let the phase angle of the phasor computed at time L be denoted by ψL then

$$\Psi_L = \Psi_{L-1} + \Delta\omega\Delta t \tag{8.6}$$

and since the time between samples is Δt seconds, the angular velocity of ψ is given by

$$\frac{d\Psi}{dt} = \frac{\Psi_L - \Psi_{L-1}}{\Delta t} = \Delta\omega \tag{8.7}$$

In other words, the computed phasor rotates in the complex plane at a rate directly related to difference between the actual power system frequency and the nominal. With $\Delta f = 1$ Hz the phasor rotates counterclockwise in the plane at a rate of one revolution per second. If $\Delta f = -1$ Hz, the rotation is clockwise. The effect is very similar to a power system synchroscope. It is important to realize that this calculation is possible because the sampling frequency is fixed at K samples per cycle of the nominal power system frequency.

The technique suggested by Equations (8.5) and (8.7) has much to recommend it for the measurement of frequency and rate of change of frequency. It uses all three phase voltages and is therefore less sensitive to error terms than techniques based on a single phase. It uses much more information than methods based on zero crossing times and is more immune to noise and harmonics. In fact, if the harmonics are multiples of the nominal power system frequency they are completely rejected. Lastly, the positive sequence voltage is a natural quantity to use in describing the system. The model for static state estimation (Section 8.4) is, in fact, the positive sequence network.

The implementation of a frequency measurement scheme using these ideas is relatively simple if it is recognized that it must take longer to measure small frequency deviations than to measure large deviations. If we wait for the rotating phasor to sweep out a reasonable angle before computing the frequency, it is clear that small frequency deviations will take longer and large deviations will be measured more quickly. A reasonable angle was found to be approximately 0.5 radians.[1] The time it would take a phasor rotating at Δf Hz to move through 0.5 radians is

$$T = \frac{0.5}{2\pi\Delta f} \text{ seconds}$$

If we assume that at least four raw measurements must be processed in order to provide smoothing then the time necessary to obtain a frequency reading for a frequency deviation of Δf Hz is

$$T = \frac{0.32}{\Delta f} \text{ seconds}$$

Thus a frequency deviation of 0.1 Hz would take 3.2 seconds while a 1 Hz deviation would take 0.32 seconds. In practice, simple averaging of the raw frequency measurements was found to be adequate to smooth the estimates.

8.2.1   Least squares estimation of f and df/dt

A more formal process for measuring frequency and rate of change of frequency is to use a least squares estimation based upon certain number of phase angles of the positive sequence measurements. Assuming that the vector of n phase angle measurements is $[\phi]$ and it is assumed that the phase angle is a quadratic function of time:

$$\phi = a_0 + a_1 t + a_2 t^2$$

If the phase angles are obtained at intervals of Δt, one may write an over-determined set of equations for solving for the coefficients a_0, a_1, and a_2:

$$\begin{bmatrix} \phi_0 \\ \phi_1 \\ \cdot \\ \cdot \\ \phi_{n-1} \end{bmatrix} = \begin{bmatrix} 1 & 0 & 0 \\ 1 & \Delta t & \Delta t^2 \\ & \cdot & \cdot \\ 1 & (n-1)\Delta t & (n-1)^2\Delta t^2 \end{bmatrix} \begin{bmatrix} a_0 \\ a_1 \\ a_2 \end{bmatrix}$$

If the coefficient matrix in the above equation is denoted by [**T**], and the unknown three-vector of 'a' coefficients is denoted by [**A**], the least squares estimate of [**A**] is

$$[\mathbf{A}] = [\mathbf{T^t T}]^{-1} \mathbf{T^t} [\phi]$$

Having calculated the coefficients, the frequency and rate of change of frequency can be obtained by direct differentiation of the assumed polynomial expression for ϕ:

$$f = f_0 + \Delta f = f_0 + (1/2\pi)(a_1 + 2a_2 t)$$

and

$$df/dt = (1/2\pi)(2a_2)$$

f_0 being the nominal system frequency. If one takes a sufficient number of phase angle samples (for example, over two or three periods of the nominal power system frequency), the above estimates are found to be very accurate.

The frequency measurement scheme has obvious application in load shedding relays, for example. It imposes almost no additional computational burden on the relaying computer and can be regarded as a by-product of the SCDFT calculation. Coupled with another development (the next section) it forms the basis for the use of measurements taken from the digital protection system in system wide control and monitoring.

8.3   Sampling clock synchronization

It has been mentioned that in an integrated substation protection system it is assumed that all sampling of voltages and currents would be synchronized. This has the advantage of allowing data sharing between modules in a backup mode, as discussed in Section 6.8. The voltage signals used in transformer protection might be obtained from line protection modules in such a system. The sampling clock synchronization would have another advantage, viz. all the phasors computed in the substation would be on a common reference. If we examine the recursive form for the DFT calculations in Equations (5.37) and (5.38), for example, it is clear that, if all the recursive calculations are begun at the same instant and updated synchronously, the phasors would all be with respect to the same reference angle. The fact that this reference angle has no particular physical meaning is unimportant. The angle differences between any two phasors computed with respect to this reference would be correct.

The obvious extension of these ideas is to synchronize sampling throughout the system. Synchronizing to an accuracy of 1 ms, in order to time-tag data, has been reported.[2] Experiments with common-reference phase angle measurements using zero-crossing instants and radio receivers tuned to standard time broadcast systems have also been discussed.[3-5] The required accuracy of synchronization can be determined by observing that at a 60 Hz power system frequency a timing error of 1μ sec. corresponds to an angular error of 0.0216 degrees (at 50 Hz it is 0.018 degrees). It would seem that sampling accuracies in the neighborhood of 1 to 10μ sec would be acceptable for almost all applications. Since the local oscillators that supervise the timing are inherently very stable (10 ppm, or less inaccuracy), an interval of one second between synchronizations is reasonable.

The other important factors that determine the choice of a synchronizing system are the cost and uninterrupted operation of the system over a long time span. There are basically two groups of systems, depending on the communication medium used: satellite and radio broadcasting systems, or fiber optic link based systems. The use of NAVSTAR satellite system to produce accuracies of a microsecond has been reported.[6] The NAVSTAR-based system has a receiver which is given the latitude, longitude, and altitude of the substation. The signal received from the satellite is processed to produce a synch pulse once a second which is accurate to within one microsecond. The system also provides the time stamp through high speed RS232C serial links (9600 Baud).

The use of fiber optic links is also an excellent choice if the high cost of implementation is not an obstacle. A fiber optic network is a good choice when the hardware is already in the system for some other purpose and the extension of its use is feasible. It should be noted that a dedicated fiber must be used for the synchronizing pulse since multiplexing will destroy the time accuracy of the snychronizing pulses. In the fiber system a master clock transmits the synch pulses with time and reference state information through a fiber optic network to all measuring systems. The fixed communication delays through the fiber system must be accounted for in the measuring systems. In either system some technique must be provided to guarantee that there are the correct number of samples between synch pulses. A phase locked loop is suggested in the reference.[6]

8.4 Application of phasor measurements to state estimation

Real-time operation of the bulk power system has been greatly enhanced by the introduction of *state estimation* algorithms. State estimation algorithms use measurements of available quantities such as: real and reactive power flows, real and reactive power injections, bus voltage magnitudes, and breaker status to estimate the *state* of the system. The state is taken to be the set of bus voltage angles and magnitudes. The estimated states are used in Energy Control Centers to determine the appropriateness of operating regimes, including the impact of contingencies, and to plan and take corrective action. The optimal power flow,[7,8] for example, assumes that the current state of the system is known from a state estimator. The transducers that make the measurements used in state estimation are called *remote terminal units* (RTUs). The RTUs communicate with the supervisory control and data acquisition system (SCADA) over dedicated telephone lines and/or microwave channels.

The commonly accepted state estimation algorithms are a form of the weighted least squares (WLS) technique of Chapter 3.[9] If the state of the system is taken as the vector $\mathbf{V}$ of bus voltage magnitudes, and the $\boldsymbol{\theta}$ vector of bus voltage angles, then the measurement vector, z, can be written as

$$\mathbf{z} = \mathbf{h}(\mathbf{V}, \boldsymbol{\theta}) + \boldsymbol{\varepsilon} \tag{8.8}$$

The measurement in Equation (8.8) is nonlinear, since the line flows and injections are nonlinear functions of the states. For example, the entry in h corresponding to the measurement of the real power flow on a line connecting bus k to bus m is given by

$$P_{km} = V_k^2 g_{km} - V_k V_m \cos\theta_{km} + V_k V_m \sin\theta_{km}$$

where the line admittance is $g_{jm} + jb_{km}$ and $\theta_{km} = \theta_k - \theta_m$. The estimator is termed *static* because the entire set of measurements z are assumed to be taken from the system in a fixed (static) state. The measurement error vector, $\boldsymbol{\varepsilon}$, is assumed to be independent from component to component with a diagonal covariance matrix given by

$$E\{\boldsymbol{\varepsilon}\boldsymbol{\varepsilon}^T\} = \mathbf{W} \tag{8.9}$$

A model of the diagonal entries in $\mathbf{W}$ is given[10] as

$$w_{ii} = \sigma_i^2$$

$$\sigma_i = (0.02m + 0.0052f_s)/3$$

where

$$m = \begin{cases} \sqrt{P_{mk}^2 + Q_{mk}^2} & \text{for flow measurements} \\ \sqrt{P_k^2 + Q_k^2} & \text{for injection measurements} \\ V_k & \text{for voltage measurements} \end{cases} \tag{8.10}$$

and

$$f_s = \text{full scale value of the instrumentation}$$

These error models are appropriate to the early SCADA systems but are inappropriate when PMU measurements are considered. The concept of meter calibration has been investigated using PMU measurements.[11–13]

Given any appropriate $\mathbf{W}$ the basic algorithm is to minimize the scalar performance index

$$J(\mathbf{V}, \boldsymbol{\theta}) = [\mathbf{z} - \mathbf{h}(\mathbf{V}, \boldsymbol{\theta})]^T \mathbf{W}^{-1}[\mathbf{z} - \mathbf{h}(\mathbf{V}, \boldsymbol{\theta})] \tag{8.11}$$

by choice of $\mathbf{V}$ and $\boldsymbol{\theta}$ (refer to Section 3.7 and equation (3.77)). Since Equation (8.11) is nonlinear it must be solved recursively. One approach is to make a linear approximation to $h(\mathbf{V},\boldsymbol{\theta})$ at the kth iteration in the form

$$h(\mathbf{V}, \boldsymbol{\theta}) = h(\mathbf{V}^k, \boldsymbol{\theta}^k) + \mathbf{H} \begin{bmatrix} \mathbf{V} - \mathbf{V}^k \\ \boldsymbol{\theta} - \boldsymbol{\theta}^k \end{bmatrix}$$

where $\mathbf{H}$ is a matrix of first partial derivatives of the components of $\mathbf{h}$ with respect to the components of $\mathbf{V}$ and $\boldsymbol{\theta}$ evaluated at $\mathbf{V}^k$ and $\boldsymbol{\theta}^k$, and the superscript k stands for the value at the kth iteration. If we write

$$\Delta \mathbf{V} = \mathbf{V} - \mathbf{V}^k$$

$$\Delta \boldsymbol{\theta} = \boldsymbol{\theta} - \boldsymbol{\theta}^k \tag{8.12}$$

$$\Delta \mathbf{z} = \mathbf{z} - h(\mathbf{V}^k, \boldsymbol{\theta}^k)$$

then the solution for $\Delta \mathbf{V}$ and $\Delta \boldsymbol{\theta}$ can be obtained from Equation (3.77), viz.

$$(\mathbf{H}^T \mathbf{W}^{-1} \mathbf{H}) \begin{bmatrix} \Delta \mathbf{V} \\ \Delta \boldsymbol{\theta} \end{bmatrix} = \mathbf{H}^T \mathbf{W}^{-1} \Delta \mathbf{z} \tag{8.13}$$

The gain matrix $\mathbf{G} = (\mathbf{H}^T \mathbf{W}^{-1} \mathbf{H})$ is not inverted as it was in Equation (3.77) because Equation (8.13) represents a large sparse set of equations. Equation (8.13) will be solved using sparse matrix techniques (factoring the matrix $\mathbf{G}$). Since the gain matrix, $\mathbf{G}$, would have to be formed and factored at each step, approximations which use a constant and decoupled form of the gain matrix are used. The decoupling is achieved by ordering the measurements into real and reactive parts in the form

$$\mathbf{z} = \begin{bmatrix} \mathbf{z}_A \\ \mathbf{z}_R \end{bmatrix}$$

where

$$\mathbf{z}_A = \begin{bmatrix} \mathbf{P}_{km} \\ \mathbf{P}_k \end{bmatrix} \quad \begin{array}{l} \text{Active Flow Measurements} \\ \text{Active Injections} \end{array} \tag{8.14}$$

and

$$\mathbf{z}_R = \begin{bmatrix} \mathbf{Q}_{km} \\ \mathbf{Q}_k \\ \mathbf{V}_k \end{bmatrix} \quad \begin{array}{l} \text{Reactive Flow Measurements} \\ \text{Reactive Injections} \\ \text{Voltage Magnitudes} \end{array} \tag{8.15}$$

so that Equation (8.13) becomes

$$\begin{bmatrix} \mathbf{G}_{AA} & 0 \\ 0 & \mathbf{G}_{RR} \end{bmatrix} \begin{bmatrix} \Delta \mathbf{V} \\ \Delta \boldsymbol{\theta} \end{bmatrix} = \begin{bmatrix} \Delta \mathbf{z}_A \\ \Delta \mathbf{z}_R \end{bmatrix} \tag{8.16}$$

The gain matrix is typically computed for $\mathbf{V} = \mathbf{1}$, and $\boldsymbol{\theta} = \mathbf{0}$. A variety of other simplifying assumptions can also be employed.[9]

One of the more important functions of a state estimator is to reject bad data. The techniques employed involve the use of $\mathbf{J}(\mathbf{V}, \boldsymbol{\theta})$ and the individual residuals Δz_i evaluated at the solution.[14,15] It is clear that if the error terms are as described by Equation (8.9) then the residuals must have a specific statistical description. If the residuals are normalized by their appropriate standard deviations, then statistical tests can be used to determine whether some residuals are suspiciously high. If the estimation is repeated with these suspect measurements removed, and the remaining residuals are acceptable, then the bad data has been identified.

Typically, only a portion of the transmission network (the HV system with voltages greater than 132 kV) are monitored. While it is desirable from a system security point of view to estimate the state of the subtransmission system, extending the estimator to lower voltages would significantly increase the computational and communication burden. The use of the measurements made by digital relaying devices with synchronized sampling clocks has been suggested as a technique of extending the estimator.[16−18]

8.4.1 WLS estimator involving angle measurements

A first possibility is to include the angle measurements directly in the WLS algorithm.[16] If some voltage angles are included the measurement vector can be modified to

$$\mathbf{z} = \begin{bmatrix} \mathbf{z_A} \\ \mathbf{z_R} \end{bmatrix}$$

where

$$\mathbf{z_A} = \begin{bmatrix} \mathbf{P_{km}} \\ \mathbf{P_k} \\ \boldsymbol{\theta_k} \end{bmatrix} \quad \begin{matrix} \text{Active Flow Measurements} \\ \text{Active Injections} \\ \text{Direct Angle Measurements} \end{matrix} \qquad (8.17)$$

and

$$\mathbf{z_R} = \begin{bmatrix} \mathbf{Q_{km}} \\ \mathbf{Q_k} \\ \mathbf{V_k} \end{bmatrix} \quad \begin{matrix} \text{Reactive Flow Measurements} \\ \text{Reactive Injections} \\ \text{Voltage Magnitudes} \end{matrix} \qquad (8.18)$$

The addition of the angle measurements to the active measurements has created an obvious symmetry to the measurement set, with the angles in active measurements playing the same role as the voltage magnitudes in the reactive measurements. The concept of including the angle measurements is not a new idea (see the discussion of the reference[17]) but is not practical without the synchronized sampling of the digital relays. The modifications to the WLS algorithm is quite simple. The performance of the estimator on the IEEE 118 bus system for different numbers of angle measurements of different qualities has been reported.[16]

There is an issue in dealing with the reference angle of the phasor measurements (see the discussion and closure of the reference[16]). In the conventional WLS algorithm one of the buses is chosen as the reference. The WLS estimates of the bus voltage angles are with respect to this chosen reference. With synchronized sampling the direct angle measurements are all with respect to a different (and non-physical) reference which is determined by the instant sampling is initiated. If the direct angle measurements are used without dealing with the reference problem unreasonable results are to be expected. One possible solution is to install a phase measuring device at the reference of the WLS system. The angle measured at the reference then should be subtracted from all the other direct angle measurements. There is a limitation involved in this solution since a failure of this single measurement would make all the direct measurements useless. We will return to the issue of a common reference after considering a second estimation technique using direct phasor measurements.

8.4.2 Linear state estimator

A more striking change in the state estimator is produced if only phasor measurements are included in the measurements set.[17] If we assume that real-time phasor measurements are made of all positive sequence bus voltages and some positive sequence currents in transmission lines and transformers, then the measurement vector is given by

$$z = \begin{bmatrix} \mathbf{V_B} \\ \mathbf{I_L} \end{bmatrix} + \begin{bmatrix} \boldsymbol{\varepsilon_B} \\ \boldsymbol{\varepsilon_L} \end{bmatrix} \tag{8.19}$$

where $\mathbf{V_B}$ and $\mathbf{I_L}$ are the true values of the bus voltages and selected line currents and $\boldsymbol{\varepsilon_B}$ and $\boldsymbol{\varepsilon_L}$ are the errors in the measurements. It should be noted that the measurement set contains exactly those quantities that would be produced by line and transformer relays. While the voltage measurements alone are sufficient to estimate the state (simply accept the measurement as the estimate) the currents are included for redundancy, to detect and identify bad data, etc. If we examine an entry I_{pq} of the vector $\mathbf{I_L}$ corresponding to a current measurement at terminal p of an element connecting nodes p and q, as shown in Figure 8.1, we can write

$$I_{pq} = y_{pq}(V_p - V_q) = y_{po}V_p \tag{8.20}$$

where y_{pq} and y_{po} are the series and shunt admittances of the element. In general then

$$\mathbf{I_L} = [\mathbf{y}\mathbf{A}^T + \mathbf{y_s}]\mathbf{V_B} \tag{8.21}$$

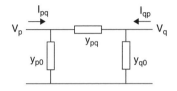

Figure 8.1   Circuit elements connecting nodes p and q

where **A** is the current measurement-bus incidence matrix (defined much as the element-bus incidence matrix[16]), **y** is a diagonal primitive admittance matrix of all the series admittances of the metered lines or transformers, and $\mathbf{y_S}$ is the diagonal primitive admittance matrix of all the shunt admittances of the metered elements at the metered ends. If the total number of buses (excluding the reference) is n, the total number of elements is b, and the number of current measurements is m, then **y** is m $\times$ m, **A** is m $\times$ n and y_S is m $\times$ n. If the current at each end of every element were measured then m = 2b. It is assumed, however, that m $\ll$ 2b. The errors are assumed to be zero mean and have covariance given by

$$\mathbf{E}\left\{\boldsymbol{\varepsilon\varepsilon}^{\mathbf{T}}\right\} = \mathbf{W} = \begin{bmatrix} \mathbf{W_B} & \mathbf{0} \\ \mathbf{0} & \mathbf{W_L} \end{bmatrix} \tag{8.22}$$

Substituting Equation (8.21) into (8.19)

$$\mathbf{z} = \begin{bmatrix} \mathbf{I} \\ \mathbf{yA^T} + \mathbf{y_s} \end{bmatrix} \mathbf{V_B} + \begin{bmatrix} \boldsymbol{\varepsilon_B} \\ \boldsymbol{\varepsilon_L} \end{bmatrix} \tag{8.23}$$

or

$$\mathbf{z} = \mathbf{B}\,\mathbf{V_B} + \boldsymbol{\varepsilon} \tag{8.24}$$

Even though **z**, **B**, and $\mathbf{V_B}$ are complex, Equation (8.24) can be solved with the least square technique. That is, following Section 3.7, we minimize the real quantity

$$(\mathbf{z} - \mathbf{B}\,\mathbf{V_B})^{\dagger}\mathbf{W}^{-1}(\mathbf{z} - \mathbf{B}\,\mathbf{V_B})$$

where $\dagger$ denotes the complex conjugate transpose of the array. The minimum is given by the solution of

$$\mathbf{B}^{\dagger}\mathbf{W}^{-1}\mathbf{B}\,\mathbf{V_B} = \mathbf{B}^{\dagger}\mathbf{W}^{-1}\mathbf{z} \tag{8.25}$$

or

$$\mathbf{G}\,\mathbf{V_B} = \mathbf{B}^{\dagger}\mathbf{W}^{-1}\mathbf{z} \tag{8.26}$$

where **G** is a gain matrix similar to that in Equation (8.13). The algorithm consists of forming the matrix **G** and finding its LU factors (the factors can be stored and used until the network structure changes or the error description changes), computing the right hand side of Equation (8.26) for each data scan, and solving Equation (8.26). Note that no iteration is involved – the system and the measurements are linear.

The structure of the **G** matrix is important in determining the practicality of the algorithm. If we write **G** in terms of the partitions in the product

$$G = [I|Ay^* + y_s^*] \begin{bmatrix} W_B^{-1} & 0 \\ 0 & W_L^{-1} \end{bmatrix} \begin{bmatrix} I \\ yA^T + y_s \end{bmatrix}$$

$$G = W_B^{-1} + y_s^* W_L^{-1} y_s + Ay^* W_L^{-1} yA^T \tag{8.27}$$

$$G = F + (H + H^\dagger) \tag{8.28}$$

where

$$F = W_B^{-1} + y_s^* W_L^{-1} y_s + Ay^* W_L^{-1} yA^T \tag{8.29}$$

and

$$H = y_s^* W_L^{-1} yA^T \tag{8.30}$$

The first and second terms in **F** are real diagonal matrices (if **W** is diagonal- which is our assumption), while the last term has the structure of an admittance matrix for a resistive network. The primitive conductance matrix is a real diagonal matrix. Indeed, **F** can be viewed as a conductance matrix for a system which has the topology of the power system but whose elements are those shown in Figure 8.2. Figure 8.2 is drawn for the element connecting nodes p and q. If only the voltages are measured the circuit of Figure 8.2(a) applies since only the first term of **F** would make a contribution at these buses. If current is measured at terminal p, the circuit of Figure 8.2(b) applies, and if current is measured at both terminals the circuit in Figure 8.2(c) results. The shunt conductances in Figure 8.2 are from the first two terms in Equation (8.29) while the series terms are from the last term in Equation (8.29). In all cases **F** is real, sparse, and symmetric, independent of the X/R ratios of the lines.

The remaining terms in **G** are from the matrix **H**. By direct calculation, if current is measured at terminal p of the line connecting terminals p and q

$$H_{pp} = y_{po}^* y_{pq}/W_{pL} \tag{8.31}$$

and

$$H_{pq} = -y_{po}^* y_{pq}/W_{pL} \tag{8.32}$$

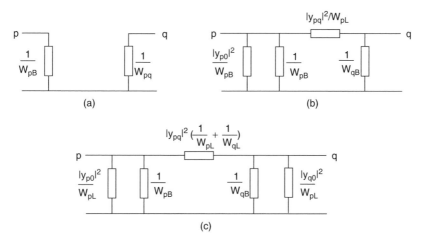

Figure 8.2 Conductance of the network represented by matrix F. The labeled values are conductances. (a) No current measurements at either end. (b) Current measurement at terminal p. (c) Current measurements at both ends

and if current is also measured at terminal q

$$H_{qq} = y^*_{qo} y_{pq}/W_{qL} \tag{8.33}$$

and

$$H_{qp} = -y^*_{qo} y_{pq}/W_{qL} \tag{8.34}$$

Thus for current measurements at both buses p and q

$$(H + H^{\dagger})_{pp} = 2\mathrm{Re}\left\{y^*_{po} y_{pq}/W_{pL}\right\} \tag{8.35}$$

$$(H + H^{\dagger})_{pq} = -y^*_{po} y_{pq}/W_{pL} - y_{qo} y^*_{pq}/W_{qL} \tag{8.36}$$

$$(H + H^{\dagger})_{qp} = -y_{po} y^*_{pq}/W_{pL} - y^*_{qo} y_{pq}/W_{qL} \tag{8.37}$$

$$(H + H^{\dagger})_{qq} = 2\mathrm{Re}\left\{y^*_{po} y_{pq}/W_{qL}\right\} \tag{8.38}$$

If the element connecting p and q is a transmission line then $y_{po} = y_{qo}$. If, in addition, the variance of the two measurements is the same ($W_{pL} = W_{qL}$) then $(\mathbf{H} + \mathbf{H}\dagger)$ is a real symmetric matrix with admittance like structure. If the element connecting p and q is a transformer with off-nominal turns ratio but with no losses, $(\mathbf{H} + \mathbf{H}\dagger)$ is once again real and symmetric.

If only one current is measured, at terminal p for example, then

$$(\mathbf{H} + \mathbf{H}^\dagger)_{pp} = 2\mathrm{Re}\left\{y^*_{po}y_{pq}/W_{pL}\right\} \tag{8.39}$$

$$(\mathbf{H} + \mathbf{H}^\dagger)_{pq} = -y^*_{po}y_{pq}/W_{pL} \tag{8.40}$$

$$(\mathbf{H} + \mathbf{H}^\dagger)_{qp} = -y_{po}y^*_{pq}/W_{pL} \tag{8.41}$$

$$(\mathbf{H} + \mathbf{H}^\dagger)_{qq} = 0 \tag{8.42}$$

In this case the matrix $(\mathbf{H} + \mathbf{H}^\dagger)$ is complex with rather small imaginary parts if the X/R ratios of the lines are large. In general, the matrix $\mathbf{G}$ is a constant – with symmetric current measurements it is real, sparse and symmetric and, if currents are measured only at one end of some elements, it has complex off-diagonal entries. Various techniques of avoiding the complex arithmetic involved with a complex $\mathbf{G}$ including the use of pseudo current measurements have been investigated.[16] The reference angle problem does not exist for the linear state estimator (LSE) since the reference is that produced by the synchronizing of the samples.

8.4.3 Partitioned state estimation

A partitioned state estimator would be one in which a portion of the network had a measurement set consisting of complex flows and injections and where the conventional WLS estimator was used, and a second distinct region where current and voltage phasor measurements were made and the LSE used. Considering the probable application, the WLS region will be referred to as the high voltage system and the LSE region will be referred to as the low voltage system. A representation of such a system is shown in Figure 8.3. Rather than approach the problem as one large WLS system it has been proposed that the two types of estimators can be combined in a way that solves the angle reference problem mentioned earlier.[18] The basic problems in combining the two estimators are the treatment of the boundary buses between the two systems and the question of reference angles for the two systems.[19–21]

The boundary buses in question are taken as the high sides of transformers connecting to the low voltage network. These buses are assumed to be included in both estimators and, since the low voltage system is assumed to be protected digitally, the phasor measurements of the boundary buses are available to both estimators. These phasor measurements can be included in the WLS estimator as in Equation (8.17) or ignored as desired. In either case, the WLS estimator is run, and produces estimates of the boundary bus angles with a reference taken as one of the high voltage

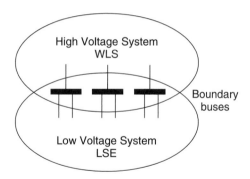

Figure 8.3   Partitioned network. The boundary buses are the high side of the transformers connecting the two systems

buses. Measurements of these same angles with the time synchronizing reference are also available. The average difference between these angles (averaged over all the boundary buses) is an accurate estimate of the difference in angle between the two references. That is,

$$\phi = \frac{1}{NB} \sum_{i=1}^{NB} (\hat{\phi}_{Hv,i} - \phi_{Lv,i}) \tag{8.43}$$

where NB is the number of boundary buses, $\theta_{Hv,i}$ is the estimate of the ith boundary bus angle form the WLS estimator and $\theta_{Lv,i}$ is the measurement of the ith boundary bus angle. The angle ϕ can be subtracted from the estimated boundary bus angles and from the angles of the estimated currents at the boundary produced by the WLS and these quantities used as measurements in the LSE algorithm. The variance of the estimate in Equation (8.43) could be further reduced by using the estimated angles from the low voltage system rather than the measurements. This would mean running the LSE algorithm twice, however. In either case, the estimate in Equation (8.43) is highly accurate if there are a number of boundary buses. Simulation results have been given for the partitioned estimator performance on the IEEE 118 bus system[20] where the WLS algorithm was used on the 345 kV system and the LSE algorithm was used on the 138 kV system.

If a sufficient number of currents are measured the LSE algorithm has bad data detection and identification properties comparable to WLS algorithms. In addition, since a number of relaying devices make phasor measurements in the substation, it is possible to pre-process data in the substation and reject bad data at a lower level. A technique of voting, averaging, or setting flags for suspect data has been described.[15] The need for all of these various approaches can be understood by considering the case of three voltage measurements made in a single substation. If all three agree within some tolerance the three could be averaged and transmitted. If one of the three is very different from the other two, the average of the two similar measurements could be transmitted to the control center. Finally, by considering four voltage measurements grouped into two agreeing groups of two each, the need for flagged data is seen. In this case the two averages would both be transmitted but with flags indicating that at least one of the pair is 'bad data'. The flags then provide guidance to the bad data detection and identification portion of the estimator. Just as in the WLS algorithm, conservation laws can be used at the substation to provide consistency checks of flow measurements. In the LSE algorithm this is simply the Kirchhoff current law. Combining the KCL check and the voltage checks approximately 75% of the bad data produced in Monte Carlo simulation was detected at the substation level.[15]

8.4.4   PMU locations

The selection of the location of PMUs has become an active research problem.[19] A number of factors motivate the interest in optimum locations for the incremental addition of PMUs to a system. Even if the SCADA system were ultimately to be replaced, it could not be done rapidly. The expense of hundreds or even thousands of PMUs must be spread out over time. For some applications the number of possible locations of PMUs may be limited. The use of PMUs in improved Power System Stabilizers, or in joining adjacent state estimators are such examples. But the state estimation problem involves virtually all power system buses. The first important observation is that if line currents are measured, then the bus voltage measurement can be extended from the PMU location to every bus that is connected to that bus. Using a variety of approaches, multiple authors have concluded that only approximately one third of the system buses need to have PMUs in order to learn all the bus voltages.[22–24] A third of the buses is still an imposing number of PMU installations so attempts to phase in the PMUs have also been attempted. Slightly different answers are obtained by different authors in these problems due to the varying amounts of preprocessing on the problem. The extremes range from solving the problem on the original intact network to doing extensive network reductions before placing PMUs.

The above is concerned with placing PMUs for state estimation. Because a large number of PMUs are ultimately required the sequential solution of the problem is an issue. Suppose the plan was to install 20 PMUs a year for five years. It would be most desirable if the best 20 locations were used in the first year and were a subset of the best 100. It is more likely that the first 20 must be the best 20 of the best 100 but not necessarily the best 20. This is an interesting problem and different than trying to find the best location for a reasonably small number of PMUs to solve a control problem. The decision trees of Section 4.10.2 have been used to choose

PMU locations for control.[25] The trees' ability to select a few variables for the splitting nodes is an attractive way to chose PMU locations.

8.5 Phasor measurements in dynamic state estimation

The estimation in Section 8.4 is static in that the power system is assumed to be in a constant state while the measurements are made. Since the phasor measurements are available as often as once a cycle, it is clear that more dynamic estimation is possible using these measurements. To simply extend the static state estimation algorithms to a once-a-cycle time frame involves much more than making faster measurements, however. The communication requirements imposed by the physical size of the power system are formidable. It is clear that only selected buses could be monitored at this high rate. A use of such a selected set of real-time measurements in power system stability is given in Chapter 8 of Reference 19 (see Reference List). The local use of rapid real-time measurements, where communication is not an issue, is a real possibility. The estimation of the internal states and parameters of synchronous machines offers an obvious application of the use of such measurements.

The estimation of machine parameters and internal states from field data has been suggested.[26–29] Either no angle measurements were used,[26] or zero crossing information was used to measure angles.[27] An example of the use of measurements of phasors and local frequency in such an application[29] was based on the system shown in Figure 8.4. It consists of a synchronous generator connected to an infinite bus through a tie line. The linear models of the exciter controller, governor controller, and turbine stage are shown in Figure 8.5. A transient was caused in the steady state operation by connecting a load to the system through switch S. The model of the synchronous machine[30] assumes that the machine has one direct axis and one quadrature axis winding on the stator, one direct and one quadrature axis damper

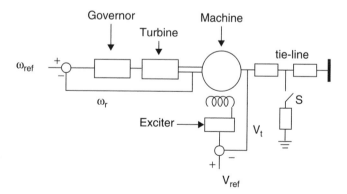

Figure 8.4   System used for dynamic estimation

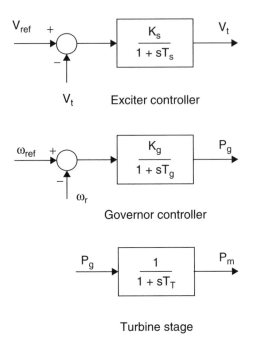

Figure 8.5 Exciter and governor controllers and turbine state of the synchronous generator

winding on the rotor and the field winding (in the direct axis) on the rotor. The flux linkages, rotor speed, and rotor angle form a convenient state for the machine. The differential equations for the synchronous generator are in the form

$$V_d = -R_a i_d - \dot{\lambda}_d - \lambda_q \omega_r$$

$$V_q = -R_a i_q - \dot{\lambda}_q - \lambda_d \omega_r$$

$$V_f = -R_f i_f - \dot{\lambda}_f \tag{8.44}$$

$$0 = -R_{kd} i_{kd} - \dot{\lambda}_{kd}$$

$$0 = -R_{kq} i_{kq} - \dot{\lambda}_{kq}$$

$$T_e = \lambda_d i_q - \lambda_q i_d$$

The flux linkages are related to the currents as follows

$$
\begin{bmatrix} \lambda_d \\ \lambda_q \\ \lambda_f \\ \lambda_{kd} \\ \lambda_{kq} \end{bmatrix}
=
\begin{bmatrix}
L & 0 & L_{af} & L_{akd} & 0 \\
0 & L_q & 0 & 0 & L_{akq} \\
L_{af} & 0 & L_f & L_{fkd} & 0 \\
L_{akd} & 0 & L_{fkd} & L_{kd} & 0 \\
0 & L_{akq} & 0 & 0 & L_{kq}
\end{bmatrix}
\begin{bmatrix} i_d \\ i_q \\ i_f \\ i_{kd} \\ i_{kq} \end{bmatrix}
\tag{8.45}
$$

where for the machine used

$$\begin{bmatrix} L_d = 0.005889 & L_{af} = 0.005335 & L_{akd} = 0.005335 \\ L_q = 0.005602 & L_{akq} = 0.005045 & L_f = 0.005615 \\ L_{fkd} = 0.005335 & L_{kd} = 0.995345 & L_{kq} = 0.005061 \\ R_a = 0.0014 & R_f = 0.0012294 & R_{kd} = 0.004349 \\ R_{kq} = 0.0132451 & & \end{bmatrix}$$

8.5.1   State equation

If we partition the state variables as follows

$$x_1^T = [\lambda_d \lambda_f \lambda_{kd} \lambda_{kq} \lambda_q V_f]$$
$$x_2^T = [P_q P_m \omega_r \delta]$$

we can write a state equation in the form

$$\begin{bmatrix} \dot{x}_1 \\ \dot{x}_2 \end{bmatrix} = \begin{bmatrix} A_{11}(\omega_r) & 0 \\ 0 & A_{22} \end{bmatrix} \begin{bmatrix} x_1 \\ x_2 \end{bmatrix} + Bu \tag{8.46}$$

where the entries in A are obtained from combining Equations (8.44), (8.45) and the numbers in Figure 8.5 (the equation is nonlinear in the state ω_r from (8.44)). The input **u** is given by

$$u^T = [V_d \quad V_q \quad \sqrt{V_d^2 + V_q^2} \; (V_d i_d + V_q i_q) \; 1] \tag{8.47}$$

At each instant of time it is assumed that the following quantities are measured

$$V_d, V_q, i_d, i_q, i_f, V_f, \omega_r, \text{ and } \delta$$

The last six are written as a measurement vector

$$z(k+1) = H \, x(k+1) + \varepsilon(k+1) \tag{8.48}$$

where ε represents the measurement errors which are assumed to be zero mean with covariance matrix Q. The nonlinear dependence of the input **u** on the measurements and the nonlinear dependence on ω_r requires the so-called extended Kalman filter,[31] which represents an approximate solution to the nonlinear filtering problem. The extended Kalman filter solution along with an observer based solution to the estimation problem has been given.[29] It should be noted that the measurements made by the SCDFT are particularly appropriate for this application – direct and quadrature components of voltages and currents are easily obtained, and the measurements of δ and ω_r are natural.

8.6 Monitoring

8.6.1 Sequence of events analysis

The ability of the integrated protection system to store samples of analog quantities and the status of contacts at every sampling instant makes it capable of providing data for sequence of events analysis. The desirability of time-tagging such information has been mentioned in the context of synchronized sampling.[2] The oscillography function of present relaying algorithms is limited by their relatively low sampling rates and the corresponding low bandwidth anti-aliasing filters required. The signals that are passed by commonly used anti-aliasing filters are limited to a few hundred Hz. The oscillographs that would be generated are unable to reproduce some system phenomena such as switching surges, traveling waves (see Chapter 9) or other high frequency components in voltages and currents. There is a substantial amount of information in the low bandwidth oscillographs, however. The low bandwidth oscillographs may be adequate for many uses.

With the steady improvement in the capability of microprocessors and A/D converters it seems possible that the broader bandwidth oscillography function may be performed by the integrated protection system in the future. It should be recognized that it is not necessary to increase the rate at which relaying calculations are performed to provide the oscillography function. One possibility is to sample at the high rate needed for oscillography (using anti-aliasing filters appropriate to the higher sampling rate) and convert the high rate samples to properly filtered low rate samples for the protection modules. A technique for this conversion along with a common data format which could be used for data exchange is presented in Appendixes III and IV.

8.6.2 Incipient fault detection

Another problem which may see some resolution with computer monitoring systems is that of detecting incipient faults in transformers. It is well known that most internal faults in a power transformer begin as small discharge currents inside the transformer tank. As these currents continue to flow, they cause further damage, accelerate the insulation breakdown, and lead to more serious permanent faults. At present, the incipient faults are detected by analyzing the gases collected in the transformer tank as a by-product of the combustion process. It may be possible to develop a technique to recognize the incipient fault condition by detecting certain features of the transformer current. It would be very difficult to detect a change in current caused by the incipient fault: it is too small compared to the transformer load current. Perhaps the frequency content of the transformer current may be a unique feature of the incipient fault. In any case, it would be necessary to suppress the steady state components from such a consideration, and also rather high sampling rates may be needed to detect the expected high frequency components in the incipient fault discharge current. Much work remains to be done in this area.

8.6.3 Breaker health monitoring

The performance of a circuit breaker can be monitored rather easily with the relays used for tripping and reclosing the breaker. One of the main concerns in circuit breaker operation is the arcing time of the main and auxiliary (if any are present) contacts. As the current in the circuit breaker is monitored by the relay, it is a simple matter to keep track of the changes in the breaker current as each contact interrupts its current. The transient monitor function discussed in Section 5.5 provides a reliable indicator of changes in the breaker current after each successful interruption. As the instant when the breaker trip coil is energized is known to the relay, rather accurate information about the breaker contact arcing can be obtained and stored in the memory of the relay. This information is of great value in determining the servicing schedule for the breaker.

By similar techniques, breaker pre-strikes can also be monitored. Since the instant of energizing the reclose coil of the breaker is known, the start of current flow in the breaker could be timed. This can be compared with the expected start delay based upon the contact travel time in its closing stroke. A premature current start indicates pre-strike, and may once again indicate necessary breaker maintenance. It is also be possible to monitor the pressure in the circuit breaker air supply system to determine if a given trip or reclose operation could be carried out successfully. In the case of insufficient air pressure, all breaker operations should be blocked, and a service alarm issued.

8.7   Control applications

A number of wide-area control applications have been described in the literature but few have found their way into the field.[19,32–37] Most are derived from linear optimal control theory and have constant feedback gains. The applications include the control of HVDC lines,[33] excitation systems,[34] power system stabilizers,[36] and FACTS devices.[37] A form of discrete control of DC lines was also proposed where DC line flows were changed discretely based on a set of phasor measurements.[25] Decision trees were used to determine the situations in which the DC line flows were changed. All of these controls were found to work effectively in simulation.

The control of low frequency inter-area oscillations with remote phasor measurements is a more developed idea. Low frequency oscillations are a growing problem in modern large-scale power systems. The oscillations are associated with under damped inter-area modes with frequencies less than one Hz. The extensive wide area measurement system and the numerous DC lines in the Chinese system have created an opportunity to address these modes with a WAMS-based control system.[38,39] The system is designed using pole placement techniques and uses measurements of angles, frequencies and real power flows. It is adaptive, in that the actual frequencies of the oscillation are measured with a fast real-time Prony algorithm[40] and the gains adjusted before the loop is closed.

8.8 Summary

This chapter has emphasized the use of the digital relay as a measuring device. The high reliability and data processing ability of the digital relay make the measuring function a logical complement to the protection function. By synchronizing sampling across the power system it is possible to obtain direct measurements of the state of the system. These real-time measurements of the bus voltage angles and magnitudes can be used to supplement or complement existing static state estimation algorithms. A particularly elegant linear estimator is produced if only real-time phasor measurements of bus voltages and line currents are used. The basic issue in connecting the two types of estimators is that of establishing a relationship between the reference angles used in each.

Real-time phasor measurements can also be used in estimating the internal states and parameters of synchronous machines from measurements at their terminals. Digital relays also can provide data for sequence of events analysis, incipient fault analysis, and breaker health. It is easy to imagine microprocessor based relays as the lowest level of a vast computer hierarchy dedicated to the integrated protection and control of the bulk power system.

Problems

8.1 Obtain the single phase version of equation $v_a(t) = \cos\omega t$. Evaluate when $\omega = 2\pi(60 + 0.1)$. Use 12 samples per cycle and a nominal frequency of 60 Hz.

8.2 Verify Equations (8.25) and (8.26). Note that z must be expressed in terms of real and imaginary parts and the objective quantity minimized with respect to each.

8.3 Write out the matrices $\mathbf{A}_{11}$, $\mathbf{A}_{22}$ and $\mathbf{B}$ in Equation (8.46).

8.4 Obtain a least squares solution to the problem of simultaneously estimating the frequency and the rate of change of frequency.

8.5 If the sampling instants have jitter, what is the influence on the frequency measurement?

8.6 If phasors at different locations have time slews, develop state estimation algorithms to estimate the constant but unknown slews.

8.7 If each sample of voltage has an independent error with zero mean and a standard deviation proportional to the sample, find the resulting error in the positive sequence voltage and in the frequency.

8.8 Given voltage phase angle measurements at both ends of a line with a synchronizing error of ΔT, how much improvement can be made in the estimate of the angle across the line by measuring the current at both ends? Assume the current phasor measurements are synchronized with the respective voltage phasor measurements.

References

[1] Phadke, A.G., Thorp, J.S., Adamiak, M.G. (1983) A new measurement technique for tracking voltage phasors, local system frequency and rate of change of frequency, IEEE Trans. on PAS., vol. 102, no. 5, pp. 1025–1038.

[2] Burnett, Jr., R.O. (1984) Field experience with absolute time synchronism between remotely located fault recorders and sequence of event recorders, IEEE Trans. on PAS, vol. 103, no. 7, pp. 1739–1742.

[3] Bonanomi, P. (1981) Phase angle measurements with synchronized clocks – principles and applications, IEEE Trans. on PAS, vol. 100, no. 11, pp. 5036–5043.

[4] Missout, G., Beland, J. and Bedard, G. (1981) Dynamic measurement of the absolute voltage angle on long transmission lines, IEEE Trans. on PAS, vol. 100, no. 11, pp. 4428–4435.

[5] Missout, G., Beland, J., Bedard, G. and Bussieres, P. (1984) Study of time dissemination methods used on electric power system with particular reference to Hydro-Quebec, IEEE Trans. on PAS, vol. 103, no. 4, pp. 861–868.

[6] Phadke, A.G., Begovic, M.M., Centeno, V.A. *et al.* (1988) Coherent sampling for system-wide digital relaying, NSF Workshop on Digital Relaying, Blacksburg, VA.

[7] Sun D.I., Ashley, B., Bewer, A. *et al.* (1984) Optimal power flow by Newton approach, IEEE Trans. on PAS, vol. 103, no. 10, pp. 2864–2880.

[8] Stott, B. and Hobson, E. (1978) Power system security control calculations using linear programming, Part I and Part II, IEEE Trans. on PAS, vol. 97, no. 5, pp. 1706–1731.

[9] Allemong, J.J., Radu, L. and Sasson, A.M. (1982) A fast and reliable state estimation algorithm for AEP's new control center, IEEE Trans. on PAS., vol. 101, no. 4, pp. 933–945.

[10] Dopazo, J.F., Ehrmann, S.T., Klitin, O.A. *et al.* (1976) Implementation of the AEP real-time monitoring system, IEEE Trans. on PAS, vol. 95, no. 5, pp. 1618–1629.

[11] Zhong, S. and Abur, A. (2005) Combined state estimation and measurement calibration, IEEE Trans. on Power Systems, vol. 20, no.1, pp. 458–465.

[12] Zhou, M. (2008) Phasor measurement unit calibration and applications in state estimation. Ph.D. Dissertation, Virginia Tech.

[13] Phadke, A.G., Thorp, J.S., Nuqui, R.F. and Zhou, M. (2009) Recent developments in state estimation with phasor measurements, IEEE/PES Power Systems Conference and Exposition (PSCE),Seattle.

[14] Handshin, E., Schweppe, F.C., Kohloas, J. and Fletcher, A. (1975) Bad data analysis for power system static state estimation, IEEE Trans. on PAS, vol. 94, no. 2, pp. 329–338.

[15] Mili, L., Van Cursem, T. and Ribbens-Pavella, M. (1985) Bad data identification meth-
     ods in power system state estimation – a comparative study, IEEE Trans. on PAS,
     vol. 104, no. 11, pp. 3037–3049.
[16] Thorp, J.S., Phadke, A.G. and Karimi, K.J. (1985) Real time voltage-phasor mea-
     surements for static state estimation, IEEE Trans. on PAS., vol. 104, no. 11,
     pp. 3098–3108.
[17] Thorp, J.S., Phadke, A.G. and Karimi, K.J. (1986) State estimation with phasor mea-
     surements, IEEE Trans. on PWRS., vol. 1, no. 1, pp. 233–241.
[18] Karimi, K.J., Thorp, J.S. and Phadke, A.G. (1986) Partitioned state estimators and bad
     data processing for static state estimators with phasor measurements, Proceedings of
     the 1986 North American Power Symposium, Ithaca, NY, pp. 131–140.
[19] Phadke A.G. and Thorp, J.S. (2008) *Synchronized Phasor Measurements and Their
     Applications*, Springer.
[20] Nuqui, R.F. and Phadke, A.G. (2005) Phasor measurement placement techniques for
     complete and incomplete observability, IEEE Trans. on Power Delivery, vol. 20, no.
     4, pp. 2381–2388.
[21] Gou, B. and Abur, A. (2001) An improved measurement placement algorithm for
     network observability, IEEE Trans. on Power Systems, vol. 16, no. 4, pp. 819–824.
[22] Abur, A. (2005) Optimal placement of phasor measurements units for state estimation,
     PSERC Publication 06-58.
[23] Rovnyak, S., Taylor, C.W. and Thorp, J.S. (1995) Real-time transient stability
     prediction – possibilities for on-line automatic database generation and classifier train-
     ing, Second IFAC Symposium on Control of Power Plants and Power Systems, Cancun,
     Mexico.
[24] Jeffers, R. (2007) Wide Area State Estimation Techniques Using Phasor Measurement
     Data. Virginia Tech Report prepared for Tennessee Valley Authority.
[25] Stagg G.W. and El-Abiad, A.H. (1968) *Computer Methods on Power Systems Analysis*,
     McGraw-Hill Company.
[26] Phadke, A.G., Thorp, J.S. and Karimi, K.J. (1986) Power system monitoring with state
     vector measurements, Second International Conference on Power System Monitoring
     and Control, Durham.
[27] de Mello, F.P., Hannet, L.N., Smith, D. and Wetzel, L. (1982) Derivation of syn-
     chronous machine parameters from pole slipping conditions, IEEE Trans. on PAS.,
     vol. 101, no. 9, pp. 3394–3402.
[28] Sugiyama, T., Nishiwaki, T., Tokedo, S. and Abe, S. (1982) Measurement of syn-
     chronous machine parameters under operating conditions, IEEE Trans. on PAS,
     vol. 101, no. 4, pp. 895–905.
[29] Namba, M., Nishiwaki, T., Yokokawa, S. *et al.* (1981) Identification of parameters for
     power system stability analysis using Kalman filter, IEEE Trans. on PAS, vol. 100,
     no. 7, pp. 3304–3311.
[30] Pillay, P., Phadke, A.G., Lindner, D.K. and Thorp, J.S. (1988) State estimation for a
     synchronous machine: Observer and Kalman filter approach, Princeton Conference.
[31] Anderson P.M. and Fouad, A.A. (1981) *Power System Control and Stability*, Iowa
     State University Press, Ames, Iowa.
[32] Gelb, A. (1975) *Applied Optimal Estimation*, MIT Press.

[33] Rostamkolai, N. (1986) Adaptive optimal control of AC/DC systems. Ph.D. Dissertation, Virginia Tech.

[34] Rostamkolai, N., Phadke, A.G., Thorp, J.S. and Long, W.F. (1988) Measurement based optimal control of high voltage AC/DC systems, IEEE Trans. on Power Systems, vol. 3 no. 3, pp. 1139–1145.

[35] Manansala, E.C. and Phadke, A.G. (1991) An optimal centralized controller with nonlinear voltage control, *Electric Machines and Power Systems* (19), pp. 139–156.

[36] Mili, L., Baldwin, T. and Phadke, A.G. (1991) Phasor measurements for voltage and transient stability monitoring and control, Workshop on Application of advanced mathematics to Power Systems, San Francisco.

[37] Snyder, A.F., Hadjsaid, N. and Georges, D. *et al.* (1998) Inter-area oscillation damping with power system stabilizers and synchronized phasor measurements, PowerCon 1998, China.

[38] Smith, M.A. (1994) Improved dynamic stability using FACTS devices with phasor measurement feedback. MS Thesis, Virginia Tech.

[39] Zhang, Z.S., Xie, X. and Wu, J. (2008) WAMS-based detection and early warning of low-frequency oscillations in large scale power systems, *Electric Power Systems Research*, 78, pp. 897–906.

[40] Shi, J.H., Li, P., Wu, X.C. *et al.* (2008) Implementation of an adaptive continuous real-time control based on WAMS, Monitoring of Power System Dynamic Performance, Saint Petersburg.

[41] Jinyu, X., Xiarong, X. *et al.* (2004) Dynamic tracking of low-frequency oscillations with improved Prony method in wide-area measurement system, Proceedings of the IEEE Power Engineering Society Meeting, vol. 1, pp. 1104–1109.

9

Relaying applications of traveling waves

9.1 Introduction

In this chapter we will discuss some developments which are technically interesting and promising. In doing so, we expose ourselves to the danger of dating this book. Only time will tell whether these concepts become widely accepted relaying practices or remain mere curiosities. On the other hand, we feel that a book such as ours is an account of this field as we view it now and therefore should take note of whatever is technically interesting and has some chance of being accepted by the industry.

We will discuss the following topics in this chapter: traveling wave relaying, differential relaying with phasors, and fault location. The developments in transducer design and communication systems brought about by fiber optic technology are also important new subjects which have been discussed in earlier chapters. These transducer and communication system advances seem certain of being widely accepted by relay engineers. It turns out that high speed communication and data processing is also an essential element of traveling wave relaying and line differential relaying. Fault location technology has clearly become an invaluable feature of most modern distance relays. It goes without saying that all three concepts are motivated and sustained by the development of computer relaying.

9.2 Traveling waves on single-phase lines

Any electrical disturbance propagates as a traveling wave on a transmission line. In single-phase transmission lines, the waves are single-mode waves – i.e. they have a single propagation velocity and characteristic impedance – whereas in a three-phase transmission line there are at least two distinct modal velocities and characteristic impedances. In either case, the occurrence of a fault sets up these traveling waves which propagate from the fault point towards the line terminals where relays are located. It is possible to design relays which utilize these propagation phenomena to detect the presence of a fault, and to determine the fault location. Since the traveling waves constitute the earliest possible evidence available to a relay that a fault has occurred, these relays have the potential of becoming the fastest responding relays. On the other hand, the traveling wave phenomena tend to contain high frequency signals (several kilohertz to megahertz, depending upon the location of the fault), and consequently the data acquisition system (A/D conversions) must have a correspondingly high bandwidth. At present, microcomputer systems would be hard-pressed to handle the traveling wave relay tasks except in cases of very long lines which produce long propagation delays. Nevertheless, the traveling wave relays represent a new and interesting development in the field of relaying, and although they may not be 'computer relays' in the manner of impedance relays discussed earlier, we will consider their operating principle in this section.

Consider a single-phase (two conductors in free space) transmission line shown in Figure 9.1. Figure 9.1(a) shows the pre-fault conditions on the line. A fault occurs at F, where the pre-fault voltage is e_F. The occurrence of a fault can be simulated by superimposing on the pre-fault network voltages and currents produced by a fault network consisting of a single source of magnitude $-e_F$ at the fault point. The relay is located at R, and is designed to sense the voltages and currents of the fault, i.e. of the network of Figure 9.1(b). In other words, the relay senses deviations in currents and voltages form their pre-fault values.

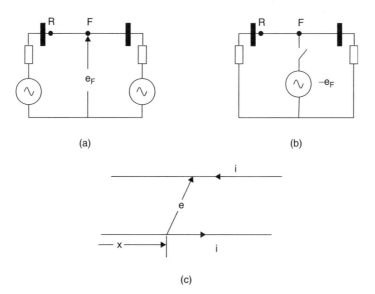

(a)   (b)

(c)

Figure 9.1   Single-phase transmission line. Wave propagation initiated by a fault. (a) Initial conditions. (b) Fault representation. (c) Wave phenomena

Now consider a point at a distance x measured from the relay location along the transmission line. The voltage and current at x obey the partial differential equations[1]

$$\frac{\partial e}{\partial x} = -L\frac{\partial i}{\partial t}$$

$$\frac{\partial i}{\partial x} = -C\frac{\partial e}{\partial t} \tag{9.1}$$

where L and C are the inductance and capacitance of the line per unit length. The resistance of the line is assumed to be negligible. Solution of Equation (9.1) is

$$e(x, t) = e_f(x - vt) + e_r(x + vt)$$

$$i(x, t) = \frac{1}{Z}e_f(x - vt) - \frac{1}{Z}e_r(x + vt) \tag{9.2}$$

where $Z = \sqrt{(L/C)}$ is the characteristic impedance of the transmission line, and $v = \sqrt{1/(L/C)}$ is the velocity of propagation.

Example 9.1

An overhead line with two conductors having equal radii of 1 cm and spacing between conductors of 10 meters has L and C given by[2]

$$L = 4 \times 10^{-7}\ln\frac{D}{0.779r} \text{ Henry/meter}$$

$$= 28.63 \times 10^{-7} \text{ Henry/meter}$$

and

$$C = n\frac{\pi \times 8.85 \times 10^{-12}}{\ln(D/r)} \text{ Farad/meter}$$

$$= 4.025 \times 10^{-12} \text{ Farad/meter}$$

For this line, the velocity of propagation

$$v = \frac{1}{\sqrt{(28.63 \times 10^{-7} \times 4.025 \times 10^{-12})}} = 2.95 \times 10^{8} \text{ m/sec}$$

and the surge impedance

$$Z = \sqrt{\frac{(28.63 \times 10^{-7})}{(4.02 \times 10^{-12})}} = 843.3 \text{ ohms}$$

It should be noted that often the line constants (L and C) of a single-phase line are expressed in 'per-phase' – i.e. between each conductor and a neutral plane half-way between the two conductors. In such a case, L becomes half and C becomes double the values computed above, the velocity of propagation remains unchanged, and the characteristic impedance becomes one-half of the value calculated above. This per-phase procedure is useful because it leads directly to the per-phase concept in a three-phase line. The Solution (9.2) of Equation (9.1) represents two traveling wavefronts: e_f traveling in the positive x direction (forward wave), and e_r traveling in the negative x direction (reverse wave). The voltage and currents at any point on the line are made up of these forward and reverse components:

$$e = e_f + e_r$$

$$i = i_f - i_r \qquad (9.3)$$

The forward and reverse components of currents are related to corresponding voltage components by the characteristic impedance Z. Figure 9.2 shows the relationships in a pictorial form for some assumed arbitrary shapes e_f and e_r. These waves propagate with a velocity v, and the net voltage and current at any point on the transmission line are given by superposition. As mentioned previously, we have neglected the resistance of the transmission lines, as well as the dependence of L upon the frequency of the voltage waves. Both of these effects are relatively minor and cause attenuation and distortion of the waveforms as they propagate along the line.[3]

Consider the occurrence of a fault as in Figure 9.2(b). The fault voltage, being a portion of the power frequency sinusoid, is approximately constant (equal to $-e_F$), and it launches two waves, both (approximately) rectangular in shape and having a magnitude of $-e_F$, moving away from the point of fault (see Figure 9.3). These waves are reflected at any discontinuity (including at the terminal where the relay is located). If k_a is the reflection coefficient applicable at terminal a, the incoming

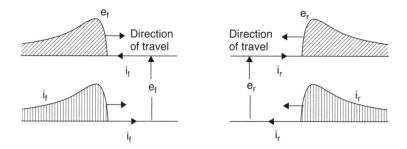

Figure 9.2   Traveling waves of voltages and currents

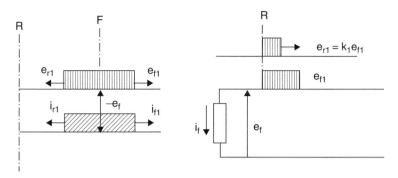

Figure 9.3 Traveling waves created by a fault. The subscript f indicates the forward traveling wave, and the subscript r indicates the reverse traveling wave with respect to the positive direction of x

wave e_{r1} causes a reflected wave moving in the forward (increasing x) direction, with its accompanying current:

$$e_{f2} = k_a e_{r1}$$
$$i_{f2} = k_a i_{r1} = e_{f2}/Z \tag{9.4}$$

The voltages and currents at the termination are (e_t, i_t), where

$$e_t = e_{r1} + e_{f2} = (1 + k_a)e_{r1}$$
$$i_t = -i_{r1} + i_{f2} = -(1 - k_a)i_{r1} \tag{9.5}$$

The value of the reflection coefficient k_a depends upon the termination: if the termination is into another identical transmission line, there are no reflections, and k_a is zero. If the termination is an open circuit, the reflection coefficient is $+1$, and at a short circuit the reflection coefficient is -1.0. For terminations into inductive or capacitive circuits, the reflection coefficient is an operator (i.e. it is a function of the Laplace variable s). If the termination is in other transmission lines, the voltage e_t and the current i_t are launched as waves on these lines, to be reflected further at their own terminations. The entire wave train, first launched by the fault, thus travels up and down the network, fragmented by the reflections, until it is dissipated through losses and the new steady state is established. When reflection coefficients are real, a very convenient method of picturing this phenomenon is through the Bewley lattice diagram.[4] Consider the circuit shown in Figure 9.4. The fault is at a distance d_a from terminal a. The propagation delays for the two segments are $t_a (= d_a/v)$ and $t_b (= d_b/v)$ respectively. At the fault point, the reflection

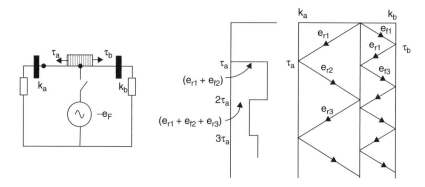

Figure 9.4   Traveling waves and the Bewley lattice diagram for a two-terminal line

coefficient is -1.0 (since the fault is assumed to be a zero impedance fault). Consequently,

$$e_{f1} = e_{r1} = -e_F$$

$$e_{f2} = k_a e_{r1}, \; e_{r2} = k_b e_{f1} \tag{9.6}$$

$$e_{r3} = -e_{f2}, \; e_{f3} = -e_{r2}$$

These successive reflections are illustrated by the lattice diagram in Figure 9.4. The voltages and currents at terminal a (as seen by a relay situated at that terminal) can be obtained by adding all the components at terminal a in the lattice diagram. For example, assuming $k_a = -0.5$, and $-e_F = 1.0$, the voltage and current waveforms at terminals a,b, and F are as shown in Figure 9.5. If the power system behind the relay location (terminal a) consists of other transmission lines with their own terminations, they in turn will have similar lattice diagrams representing reflections at their terminals. Reflections which return toward terminal a will once again impinge upon it, and produce waves propagating on the faulted line. Consider one such line a-c in Figure 9.6. Another line of infinite length is also assumed to exist in parallel with line a-c. (The length of this line is assumed to be infinite in order to simplify our discussion). The reflection coefficients at c and a for waves traveling on line a-c are calculated as before. It is clear that the resulting waveforms at a are now far more complex.

So far we have assumed that the fault voltage $-e_F$ is constant. However, in reality, it varies as a sine wave of power frequency. If a fault occurs at 100 km from the relay location, t_a is about 0.33 msec, and a 60 Hz waveform may be considered to be constant over a few travel times. For faults occurring at shorter distances, the assumption of constancy of e_F is closer to reality. In any case, it should be noted that the steady state voltages and current at the relay location are sinusoidal, with the current having a decaying DC offset whose magnitude depends upon the fault incidence angle. The current and voltage at the relay location build up according to the traveling wave considerations discussed so far and finally evolve into their respective steady state values.

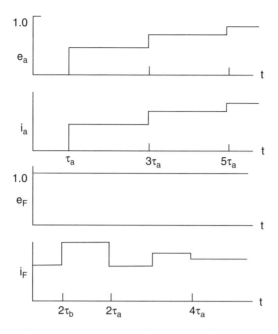

Figure 9.5   Voltage and current waveforms at line terminal and fault point due to traveling waves

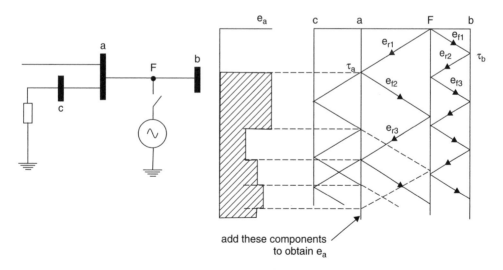

Figure 9.6   Multiple reflections of waves created by the fault at F. One line at bus a is assumed to be infinite in length for the sake of simplicity. The other line a-c is terminated in a load impedance

9.3   Traveling waves on three-phase lines

Three-phase transmission lines consist of phases a, b and c, and a ground system consisting of earth and ground wires (if they are present). As in case of single-phase transmission lines, the voltages and currents at a distance x from the line terminal are related by partial differential equations

$$\frac{\partial}{\partial x}\begin{bmatrix} e_a \\ e_b \\ e_c \end{bmatrix} = - \begin{bmatrix} L_s & L_m & L_m \\ L_m & L_s & L_m \\ L_m & L_m & L_s \end{bmatrix} \frac{\partial}{\partial x} \begin{bmatrix} i_a \\ i_b \\ i_c \end{bmatrix}$$

$$\frac{\partial}{\partial x}\begin{bmatrix} i_a \\ i_b \\ i_c \end{bmatrix} = - \begin{bmatrix} C_s & C_m & C_m \\ C_m & C_s & C_m \\ C_m & C_m & C_s \end{bmatrix} \frac{\partial}{\partial x} \begin{bmatrix} e_a \\ e_b \\ e_c \end{bmatrix} \tag{9.7}$$

In Equation (9.7) we have assumed that the transmission line is transposed. The self-inductance of each phase is L_S H/m, and the mutual inductance between any two phases is L_m H/m. Similarly, the capacitance between each phase and ground is C_S F/m, and the capacitance between any two phases is C_m. Usually the quantities L_S, L_m, C_S, C_m are expressed in terms of their positive and zero sequence values (which are eigen values of the L and C matrices respectively):[4]

$$L_s = \frac{1}{3}(L_0 + 2L_1)$$

$$L_m = \frac{1}{3}(L_0 - L_1)$$

$$C_s = \frac{1}{3}(C_0 + 2C_1) \tag{9.8}$$

$$C_m = \frac{1}{3}(C_1 - C_0)$$

The corresponding circuit representation is shown in Figure 9.7. Note that Figure 9.7(a) shows the conductor-to-conductor capacitances in a Wye connection, whereas Figure 9.7(b) shows them in a delta connection. Equations (9.6) have a traveling wave solution in modal quantities. Let 0, α ,β be the Clarke components of the voltages and currents:

$$\begin{bmatrix} i_0 \\ i_\alpha \\ i_\beta \end{bmatrix} = \frac{1}{3}\begin{bmatrix} 1 & 1 & 1 \\ 2 & -1 & -1 \\ 0 & \sqrt{3} & -\sqrt{3} \end{bmatrix}\begin{bmatrix} i_a \\ i_b \\ i_c \end{bmatrix}$$

$$\begin{bmatrix} e_0 \\ e_\alpha \\ e_\beta \end{bmatrix} = \frac{1}{3}\begin{bmatrix} 1 & 1 & 1 \\ 2 & -1 & -1 \\ 0 & \sqrt{3} & -\sqrt{3} \end{bmatrix}\begin{bmatrix} e_a \\ e_b \\ e_c \end{bmatrix} \tag{9.9}$$

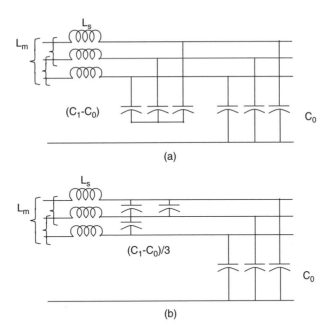

Figure 9.7   Multi-phase transmission line model. (a) Wye form. (b) Delta form

In this case, the solution of Equation (9.7) for voltages and currents at any point on the three-phase line is given by

$$e_0 = e_{f0}(x - v_0 t) + e_{r0}(x + v_0 t)$$

$$e_\alpha = e_{f\alpha}(x - v_1 t) + e_{r\alpha}(x + v_1 t) \tag{9.10}$$

$$e_\beta = e_{f\beta}(x - v_1 t) + e_{r\beta}(x + v_1 t)$$

$$i_0 = i_{f0} - i_{r0}$$

$$i_\alpha = i_{f\alpha} - i_{r\alpha} \tag{9.11}$$

$$i_\beta = i_{f\beta} - i_{r\beta}$$

where

$$\frac{e_{f0}}{i_{f0}} = \frac{e_{r0}}{i_{r0}} = Z_0 = \sqrt{\frac{L_0}{C_0}}$$

$$\frac{e_{f\alpha}}{i_{f\alpha}} = \frac{e_{r\alpha}}{i_{r\alpha}} = Z_1 = \sqrt{\frac{L_1}{C_1}} \tag{9.12}$$

$$\frac{e_{f\beta}}{i_{f\beta}} = \frac{e_{r\beta}}{i_{r\beta}} = Z_1$$

and

$$v_0 = \frac{1}{\sqrt{L_0 C_0}}$$

$$v_1 = \frac{1}{\sqrt{L_1 C_1}} \tag{9.13}$$

In other words, phase voltages and currents should be viewed as being made up of $0, \alpha, \beta$ components. Each of these components represents a mode of the propagation Equation. The 0-mode has a characteristic impedance and velocity of propagation that are distinct from those of the α, β modes.

Example 9.2

Consider the $\mathbf{L}$ and $\mathbf{C}$ matrices given below:

$$\mathbf{L} = \begin{bmatrix} 2.1 & 0.8 & 0.8 \\ 0.8 & 2.1 & 0.8 \\ 0.8 & 0.8 & 2.1 \end{bmatrix} \times 10^{-6} \, \mathrm{H/m}$$

$$\mathbf{C} = \begin{bmatrix} 8.77 & -1.033 & -1.033 \\ -1.033 & 8.77 & -1.033 \\ -1.033 & -1.033 & 8.77 \end{bmatrix} \times 10^{-12} \, \mathrm{F/m}$$

Then

$$L_0 = 3.7 \times 10^{-6} \mathrm{H/m}, \qquad L_1 = 1.3 \times 10^{-6} \, \mathrm{H/m}$$
$$C_0 = 6.704 \times 10^{-12} \mathrm{F/m}, \quad C_1 = 9.8 \times 10^{-12} \, \mathrm{F/m}$$

The zero-sequence mode parameters are

$$Z_0 = \sqrt{\frac{L_0}{C_0}} = 742.9 \, \mathrm{ohms}$$

$$v_0 = \sqrt{\frac{1}{L_0 C_0}} = 2 \times 10^8 \, \mathrm{m/sec}$$

and the α and β sequence mode parameters are

$$Z_1 = \sqrt{\frac{L_1}{C_1}} = 364.2 \, \mathrm{ohms}$$

$$v_1 = \sqrt{\frac{1}{L_1 C_1}} = 2.8 \times 10^8 \, \mathrm{m/sec}$$

9.3.1 Traveling waves due to faults[5]

Let the pre-fault voltages of phases a, b, and c be e_{aF}, e_{bF}, and e_{cF}, respectively. For a *three-phase fault*, the fault is represented by voltages $-e_{aF}$, $-e_{bF}$, $-e_{cF}$ at the fault point. From Equations (9.8), the 0,α,β components of these voltages are given by (since $e_a + e_b + e_c = 0$ at any instant)

$$e_{OF} = \frac{1}{3}(-e_{aF} - e_{bF} - e_{cF}) = 0$$

$$e_{\alpha F} = \frac{1}{3}(-2e_{aF} + e_{bF} + e_{cF}) = -e_{aF} \tag{9.14}$$

$$e_{\beta F} = \frac{1}{3}(-e_{bF} + e_{cF})$$

The α and β components of voltages are launched as forward and reverse waves of magnitude $e_{\alpha f}$, $e_{\alpha r}$, $e_{\beta f}$, $e_{\beta r}$. Their accompanying current waves are given by

$$i_{\alpha r} = \frac{e_{\alpha r}}{Z_1}, \quad i_{\alpha f} = \frac{e_{\alpha f}}{Z_1}$$

$$i_{\beta r} = \frac{e_{\beta r}}{Z_1}, \quad i_{\beta f} = \frac{e_{\beta f}}{Z_1} \tag{9.15}$$

As in the case of a single-phase transmission line, each mode (α,β) now travels up and down the line, and their travels can be catalogued by separate lattice diagrams as in Figure 9.6.

For a b-c fault, the fault is defined by the boundary condition

$$e_a = 0$$

$$e_b = -\frac{1}{2}e_{bcF} \tag{9.16}$$

$$e_c = +\frac{1}{2}e_{bcF}$$

where e_{bcF} is the pre-fault voltage between phases b and c. The above condition, when substituted in the definition of Clarke components (Equation (9.9)) leads to

$$e_0 = 0$$

$$e_\alpha = 0 \tag{9.17}$$

$$e_\beta = \frac{1}{\sqrt{3}}e_{bcF}$$

Similarly (see Problem 9.5), the modal waves launched by various types of faults can be calculated. Table 9.1 lists the waves for the ten fault types possible on a three-phase system.

Table 9.1 $0, \alpha, \beta$ waves launched by faults on a three-phase transposed transmission line

Fault type	e_0	e_α	e_β
3 ph	0	$-e_{aF}$	$-\dfrac{1}{\sqrt{3}}(e_{bF} - e_{cF})$
ab	0	$-\dfrac{1}{2}e_{abF}$	$\dfrac{1}{2\sqrt{3}}e_{abF}$
bc	0	0	$-\dfrac{1}{\sqrt{3}}e_{bcF}$
ca	0	$-\dfrac{1}{2}e_{caF}$	$\dfrac{1}{2\sqrt{3}}e_{caF}$
abg	$-\dfrac{Z_0(e_{aF} + e_{bF})}{Z_1 + 2Z_0}$	$-\dfrac{(Z_1 + Z_0)e_{aF}}{Z_1 + 2Z_0}$ $+\dfrac{Z_0 e_{bF}}{Z_1 + 2Z_0}$	$\dfrac{(Z_0 - Z_1)e_{aF}}{\sqrt{3}(Z_1 + 2Z_0)}$ $-\dfrac{(2Z_1 + Z_0)e_{bF}}{\sqrt{3}(Z_1 + 2Z_0)}$
bcg	$-\dfrac{Z_0(e_{bF} + e_{cF})}{Z_1 + 2Z_0}$	$\dfrac{Z_1(e_{bF} + e_{cF})}{Z_1 + 2Z_0}$	$-\dfrac{(e_{bF} - e_{cF})}{\sqrt{3}}$
cag	$-\dfrac{Z_0(e_{aF} + e_{cF})}{Z_1 + 2Z_0}$	$-\dfrac{(Z_1 + Z_0)e_{aF}}{Z_1 + 2Z_0}$ $+\dfrac{Z_0 e_{cF}}{Z_1 + 2Z_0}$	$\dfrac{(Z_0 - Z_1)e_{aF}}{\sqrt{3}(Z_1 + 2Z_0)}$ $-\dfrac{(2Z_1 + Z_0)e_{cF}}{\sqrt{3}(Z_1 + 2Z_0)}$
ag	$\dfrac{Z_0 e_{aF}}{Z_0 + 2Z_1}$	$-\dfrac{Z_1 0 e_{aF}}{Z_0 + 2Z_1}$	0
bg	$\dfrac{Z_0 e_{bF}}{Z_0 + 2Z_1}$	$\dfrac{Z_1 e_{bF}}{Z_0 + 2Z_1}$	$-\dfrac{\sqrt{3}Z_1 e_{bF}}{Z_0 + 2Z_1}$
cg	$\dfrac{Z_0 e_{cF}}{Z_0 + 2Z_1}$	$\dfrac{Z_1 e_{cF}}{Z_0 + 2Z_1}$	$\dfrac{\sqrt{3}Z_1 e_{cF}}{Z_0 + 2Z_1}$

Thus, every type of fault launches a set of $0, \alpha, \beta$ waves at the point of a fault inception. The waves propagate towards terminals, and are reflected with appropriate reflection coefficients. In the case of unbalanced faults, the reflections at the fault point are more complex: the wave of one type $(0, \alpha,$ or $\beta)$ in general is reflected and transmitted as waves of all three types. In other words, an unbalanced fault causes coupling between various modes of propagation. The reflection and transmission coefficient calculations are left as a problem for the reader (see Problem 9.7).

It should also be noted that, although the characteristic impedances and velocities of propagation of the three modes are determined by the $\mathbf{L}$ and $\mathbf{C}$ matrices of Equation (9.6), the definitions of the three modes (as for Clarke components) are not unique. Other possibilities exist, the most common alternative definition being the Karrenbauer transformation:[6]

$$\begin{bmatrix} e_0 \\ e_1 \\ e_2 \end{bmatrix} = \frac{1}{3} \begin{bmatrix} 1 & 1 & 1 \\ 1 & -1 & 0 \\ 1 & 0 & -1 \end{bmatrix} \begin{bmatrix} e_a \\ e_b \\ e_c \end{bmatrix} \tag{9.18}$$

9.4 Directional wave relay

As the waves created by faults travel in distinct directions (away from and towards a fault point), it is possible to design a relay which depends upon the traveling waves to determine the direction of the fault with respect to the relay location. If two relays at the ends of a line detect a fault to be in the forward direction, then the fault is in the zone of protection of both relays. This principle of relaying (directional comparison) requires a communication channel between the two ends to confirm that both relays see the fault in the forward direction.

Since a relay is located at the terminal of a transmission line, it would appear that it sees waveforms that are affected by the discontinuity represented by the bus through the reflection coefficients. As a matter of fact, the relays may be made to see a specific traveling wave in a manner that is independent of the terminations. Consider the 'discriminant functions' d_f and d_r associated with the forward and reverse waves:[6]

$$d_f = e + Zi$$
$$d_r = e - Zi \tag{9.19}$$

where e and i are the modal voltages and currents $(0, \alpha,$ or $\beta)$ at the relay location. The relay treats the currents flowing into the line as positive. The forward current wave i_f thus appears as positive to the relay, while the reverse current wave i_r appears to be negative. It follows from Equation (9.3) that

$$d_f = e + Zi = e_f + e_r + Z(i_f - i_r)$$
$$= (e_f + Zi_f) + (e_r - Zi_r) \tag{9.20}$$
$$= e_f + Zi_f = 2e_f$$

since $e_r = Zi_r$. Similarly,

$$d_r = e_r + Zi_r = 2e_r$$

Example 9.3

The discriminant functions d_f and d_r at bus a of Figure 9.8 are shown in Figure 9.9. Note that the discriminant function d_f and d_r remain constant at $2e_f$ and $2e_r$ respectively until the other reflections due to fault (at instants $t_3, t_5 \ldots$), or due to reflections at bus c (at instants $t_2, t_4 \ldots$), cause a change in their values. If the relay is not a discontinuity, only d_r picks up at t_1 and d_f must await reflections either at the fault or at bus c.

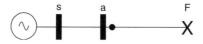

Figure 9.8 System one line diagram for traveling wave considerations

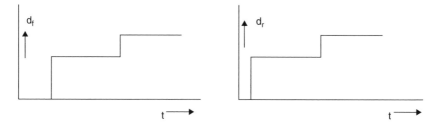

Figure 9.9 Discriminant functions for the relay at bus a for the system shown in Figure 9.8. Waves traveling towards the relay are termed reverse waves since the relay is assumed to be looking into the line

If the fault is behind the relay, d_f will pick up (i.e. become non-zero) first, and d_r will pick up later due to successive reflections at the fault or at bus c. Thus we may determine the direction of the fault according to the following logic:[7]

(i) if d_f picks up first, the fault is behind the relay

(ii) if d_r picks up first or simultaneously with d_f, the fault is in the forward direction.

The relationship (ii) above is maintained for a period equal to twice the travel time between the forward fault and the relay location. It would appear that for a near fault the duration of validity could be very brief since the travel time from the fault would be very short. However, it has been observed[6] that, for near faults, the quick successive reflections build up the discriminant functions d_f and d_r d_r still maintaining the required relationship (ii). The details are left as an exercise for the reader (see Problem 9.8).

Another concern with discriminant functions of Equation (9.20) is that, for faults that occur at near zero voltage (i.e. when the pre-fault voltage is going through a zero), e_f and e_r are both very small. The traveling waves (and therefore the discriminants) are replicas of the (negative) pre-fault voltages. As time goes on, these discriminant functions acquire the shape of a power frequency sine wave. Recall that, although a sine wave is small in the beginning, its derivative there is quite large, and one could construct modified discriminant functions[6]

$$d'_f = \sqrt{d_f^2 + \left\{ \frac{1}{\omega} \frac{d}{dt}(d_f) \right\}^2}$$

$$d'_r = \sqrt{d_r^2 + \left\{ \frac{1}{\omega} \frac{d}{dt}(d_r) \right\}^2} \qquad (9.21)$$

These primed discriminant functions are independent of the fault incidence angle.

Other versions of the traveling wave algorithm consider the fault trajectory in the (e-Ki) plane, where K is a constant. Note that a constant discriminant function given in Equation (9.19) defines a straight line in this plane, when K is set equal to the characteristic impedance of the line. A decision about the direction of a fault can be based upon whether the (e-Ki) trajectory crosses appropriate thresholds in the (e-Ki) plane.[8,9] The constant K may be set equal to Z, the source impedance of the network. In this case the current is shifted in phase by Z (the source impedance being a complex quantity),[10–12] so that e and Ki are in phase or in phase opposition with each other depending upon whether the fault is in front of or behind the relay location. Note that these latter (and some other) relays[13] are in fact power frequency relays operating on incremental quantities, rather than traveling relays. When K is made equal to Z, it is once again a mimic circuit used to suppress the DC offset in the current and produce a phase shift in the current waveform. The traveling wave components then become parasitic effects which must be ignored, and actual relaying decisions are based upon fundamental frequency voltages and currents.

9.5 Traveling wave distance relay

Consider a single-phase transmission line with a relay at bus a and a fault at bus F as shown in Figure 9.8. As explained in Section 9.2, the occurrence of the fault sets up traveling waves which travel away from the fault. The waves are reflected at all discontinuities according to the lattice diagram of Figure 9.4. If we were to find a wave that went from the relay location towards the fault and started a timer as the wave went through the relay location, the wave would be reflected by the fault and return to the relay location. When the wave crosses the relay location in the reverse direction, the timer is stopped. The timer reading would then correspond to twice the travel time needed to traverse the distance to the fault.

Since the velocity of propagation of the wave is known, the distance to the fault can be calculated. If we assume the distance sa to be about half of af in Figure 9.8,

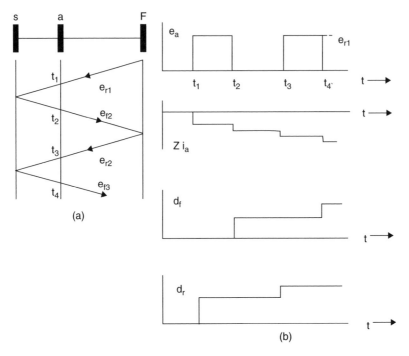

Figure 9.10 Voltage, current and discriminant functions for a fault. (a) System diagram and the lattice diagram. (b) Waveforms. The reverse discriminant function arrives at the relay location first because it is associated with the wave traveling towards the relay from the fault

a fault with zero impedance, and a source of infinite short circuit capacity at bus s, the traveling waves as seen by the relay can be obtained from the lattice diagram shown in Figure 9.10. At time t_2, a wave reflected at the source behind the relay crosses the relay and goes towards the fault. At time t_3 this wave is reflected by the fault and crosses the relay location in the reverse direction. Consequently, $(t_3 - t_2)$ is twice the travel time between the relay location and the fault. The forward and reverse discriminant functions for this case are also plotted in Figure 9.10. The relay must be allowed to start its timer at t_2 when the pickup of d_f indicates the arrival of a wave after (or simultaneously) d_r has picked up. This, as was stated in the last section, indicates a forward fault. Once the timer has started, it must be stopped when d_r shows a waveform similar to the one which crossed the relay location at t_2.

The square wave shapes shown in Figure 9.10 result from assumed zero impedance in the fault and the source. If the source is inductive, the reflected waves at the source will be exponential functions as shown in Figure 9.11. The effect of fault arc resistance (when it is present) is to increase the voltage reflection coefficient from -1.0 corresponding to a direct short-circuit to a value somewhat closer to zero depending upon the resistance in the arc.

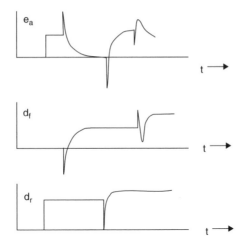

Figure 9.11   Discriminant wave functions for inductive termination

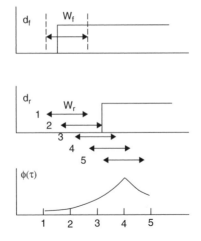

Figure 9.12   Discriminant functions and their cross-correlation functions

One could realize the starting and stopping operations of the timer by actually using a clock, in which case adequate triggering mechanisms must be built in. Alternatively,[14,15] one could store the waveform of d_f around the instant t_2 for a time window w_f spanning t_2 as shown in Figure 9.12. The cross-correlation function of d_f (taken over the window w_f) and d_r (taken over a winder w_r) of equal duration at progressively increasing time delays τ may be used as an algorithmic measure for finding the interval $(t_3 - t_2)$:

$$\varphi(\tau) = \int_{t=0}^{t=w_f} d_f(t)d_r(t+\tau)dt \qquad (9.22)$$

As has been pointed out,[14] the DC component of d_f and d_r over windows w_f and w_r produces a large (and variable) bias in $\varphi(\tau)$ which would mask the effect we are seeking. It is therefore appropriate to modify Equation (9.22) by removing the mean of d_f and d_r from the cross-correlation function. Let

$$\overline{d}_f = \frac{1}{w_f} \int\limits_{t=0}^{t=w_f} d_f(t)dt$$

$$\overline{d}_r(\tau) = \frac{1}{w_r} \int\limits_{t=0}^{t=w_r+\tau} d_f(t)dt$$

A new cross-correlation function $\varphi'\tau$ is defined

$$\varphi'\tau = \int\limits_{t=0}^{t=w_f=w_r} (d_f - \overline{d}_f) \left\{ d_r(t+\tau) - \overline{d}_r(\tau) \right\} dt \tag{9.23}$$

Both the function $\varphi'(\tau)$ and its sampled data version become maximum when τ becomes equal to $(t_3 - t_2)$. The location of the fault can now be easily determined.

Methods to find the distance to the fault from traveling waves are fraught with difficulty. In a three-phase system, multi-mode propagation exists. Ground mode waves are subject to severe attenuation and distortion of wave shape as the waves propagate along the transmission lines. Unbalanced faults and faults through impedance further complicate matters by coupling different modes at the fault point. When fault inception angles are such that traveling waves are small in magnitude, the maximum of the cross-correlation function $\varphi'(\tau)$ is often lost in the measurement noise. All these considerations make a traveling wave distance relay somewhat difficult to set. It could well be that it must be used in conjunction with other more conventional relaying schemes.

9.6 Differential relaying with phasors

Differential relaying is the preferred protection method for power apparatus. We have covered this principle in the discussion of transformer, generator, and bus protection in Chapter 6. Differential protection of cables and transmission lines is a relatively new subject, which we will consider in this section.

Differential relaying requires that information (usually currents) from all terminals of the zone being protected be combined with each other to form differential current. If this exceeds a preset value, a fault may be said to exist inside the zone of protection. Such a computation must be made at each terminal, and consequently a communication channel must be assumed to exist between all terminals of the

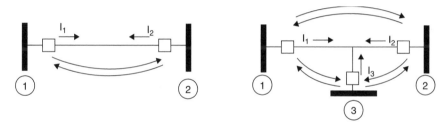

Figure 9.13 Differential relaying of a transmission line and the necessary communication paths. Either a two-terminal line as shown on the left, or a multi-terminal line as on the right, can be protected by this scheme

protected line or cable. Figure 9.13 shows the communication channels needed for differential protection of two- and multi-terminal transmission lines.

We may begin with a differential protection based upon phasors. The phasors may be calculated from fractional cycle data, as discussed in Chapter 2. If I_i is the current phasor at terminal i (reference direction is positive when the current is flowing into the zone of protection), the differential currents may be defined in the usual manner:

$$|I_d| = \left| \sum_i I_i \right| \tag{9.24}$$

A single restraining current may be constructed by averaging the magnitudes of all terminal currents, or one restraining current for every pair of terminals may be constructed in order to maintain uniform sensitivity when one of the terminals of a multi-terminal line is out of service. The reader may refer to Chapter 5, where these considerations are explained in the context of multi-winding transformer protection.

It is clear that all the currents in Equation (9.24) must be on a common reference, so that synchronized sampling clocks as described in Chapter 8 must be used. It is also possible to achieve synchronization through pre-fault load flow on the transmission lines.[16] Assume that voltages and currents at each terminal of the transmission line are measured in the period immediately before the occurrence of a fault. All phasors within a station can certainly be synchronized. Thus the phasors I_i and E_i are on the same reference, but their reference may be at a certain unknown angle δ_i with respect to a system-wide common reference. If the transmission line is represented by a bus admittance matrix Y_B as shown in Figure 9.14, the bus injection currents and voltages are related by:

$$\mathbf{I_B} = \begin{bmatrix} I_{1a}\,e^{j\delta}1 \\ I_{1a}\,e^{j\delta}1 \\ I_{1c}\,e^{j\delta}1 \\ I_{2a}\,e^{j\delta}2 \\ I_{1b}\,e^{j\delta}1 \\ I_{2c}\,e^{j\delta}2 \end{bmatrix} = \mathbf{Y_B} \begin{bmatrix} E_{1a}\,e^{j\delta}1 \\ E_{1a}\,e^{j\delta}1 \\ E_{1c}\,e^{j\delta}1 \\ E_{2a}\,e^{j\delta}2 \\ E_{2b}\,e^{j\delta}2 \\ E_{2c}\,e^{j\delta}2 \end{bmatrix} \tag{9.25}$$

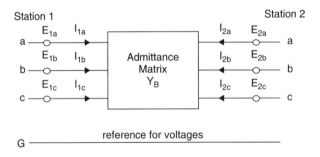

Figure 9.14 Two-terminal transmission line representation for synchronization of phasors with pre-fault load flow

δ_1 may be assumed to be zero without loss of generality. Once could thus solve Equation (9.25) as a redundant set of equations for a single unknown δ_2. For an n-terminal line, the n-1 unknown angles could be found in a similar manner. The details may be found in the literature.[16] Once the unknown angles from a common reference are determined, these angles are to be held constant for differential-current calculations during a fault period. When no transient is detected, the synchronization is carried out continuously – perhaps as frequently as once a cycle.

When charging currents are significant, the differential current is no longer zero for a no-fault condition. Thus in the case of long transmission lines or cables, the differential relay must be made insensitive in order to accommodate the charging current. An alternative method is to calculate a correction to the differential current due to the charging currents. If we assume a two-terminal multi-phase π-section representation of a transmission line, the charging currents $\mathbf{I}_{si}$ at the terminal **I** are given by

$$\mathbf{I_{si}} = \begin{bmatrix} I_{sia} \\ I_{sib} \\ I_{sic} \end{bmatrix} = \begin{bmatrix} \mathbf{Y_S} \end{bmatrix} \begin{bmatrix} E_{ia} \\ E_{ib} \\ E_{ic} \end{bmatrix} \qquad (9.26)$$

where $\mathbf{Y_S}$ is the shunt capacitive admittance matrix at bus i. Thus, instead of sending $\mathbf{I}_i$ from terminal **I** to all other terminals, a compensated current $\mathbf{I}_i'$

$$\mathbf{I_i'} = \mathbf{I_i} - \mathbf{I_{si}} \qquad (9.27)$$

is sent to the remote terminals. For no faults or external faults, the differential current calculated with $\mathbf{I}_i'$ is zero even in the presence of charging currents. A somewhat modified method for computing $\mathbf{I}_i'$ must be used when dealing with a multi-terminal transmission line, since the charging currents at the tap points must also be reckoned with. This problem is left as an exercise at the end of this chapter. Additional details may be found in the literature.[17]

It is clear that the differential protection principle described by Equations (9.26) and (9.27) requires voltage measurement. This may not always be available at every terminal. One must either abandon the idea of compensation at all terminals, or use it at those terminals where it is available. In any case, inaccuracies introduced by such approximation must be allowed for by an appropriate de-sensitization of the differential relay.

9.7 Traveling wave differential relays

As explained in earlier sections, a fault launches traveling waves, which ultimately (after successive reflections and attenuations) produce standing wave patterns (i.e. steady state conditions) on the transmission lines which are the phasor voltages and currents used in the relaying principle described in Section 9.7. It is also possible to define a differential relaying principle which may be based upon the traveling waves.[18,19]

If the travel time between two ends of a line is τ, and if at both ends of a line a positive direction for forward propagation is into the zone of protection, it is clear that the forward wave at a terminal becomes a reverse wave at the other terminal in time τ. Thus d_f at time t at terminal 1 in Figure 9.13 becomes d_r at $(t + \tau)$ at terminal 2:

$$d_{f1}(t) = d_{r2}(t + \tau) \tag{9.28}$$

Equation (9.28) holds as long as there is no fault on the transmission line. Recalling that the discriminant functions defined by Equation (9.18) may be modified to the following form:

$$d_{f1}(t) = e_1(t) + Zi_1(t) = Z\left\{i_1(t) + \frac{1}{Z}e_1(t)\right\} \tag{9.29}$$

and similarly

$$d_{r2}(t + \tau) = Z\left\{-i_2(t + \tau) + \frac{1}{Z}e_2(t + \tau)\right\} \tag{9.30}$$

leaving the common factor Z out since it is a constant, one could construct a differential current at the two terminals of a transmission line:

$$i_{d1} = \left\{i_1(t) - i_2(t + \tau) + \frac{1}{Z}e_1(t) - e_2(t + \tau)\right\} \tag{9.31}$$

$$i_{d2} = \left\{i_2(t + \tau)_2 - i_1(t)\right\} + \frac{1}{Z}\left\{e_2(t - \tau) - e_1(t)\right\} \tag{9.32}$$

If current and voltage samples at terminals 1 and 2 are taken τ seconds apart, the differential currents i_{d1} and i_{d2} can be computed simply if the samples are exchanged between the two terminals. As before, tolerances must be set for residual i_d which may exist during external faults due to errors in the measurement process. Thus a percentage differential relaying principle should be used, although the percentage slope would be very small for a well designed data acquisition system.

In a three-phase system, multi-mode propagation must be considered. Thus $0,\alpha,\beta$ differential currents must be evaluated and used separately. Usually the zero-sequence characteristic impedance and velocity of propagation are significantly different from the α and β mode parameters, as explained in Section 9.3. Consequently, samples at the two terminals would have to be taken with delays of τ_0 and τ_1, where τ_0 and τ_1 are travel times of the 0 and α,β modes respectively. This would lead to a very complex sampling process, and an alternative might be to use identical parameters for all the modes, and accept larger values of i_{d1} and i_{d2} for external faults. Yet another simplification may be to take samples at a fixed delay (say τ_1), and then use interpolation to determine samples τ_0.[18,19]

It should be recalled that the steady state (phasor) solutions of the traveling wave Equation lead to standing waves – made up of one traveling forward and another one in the reverse direction. Thus, when steady state conditions are reached, Equations (9.31) and (9.32) are equivalent to Equation (9.27). Verification of this is left as an exercise for the reader (see Problem 9.11).

9.8   Fault location

9.8.1   Impedance estimation based fault location

Although the transmission line protection algorithms discussed in Chapter 3 use estimates of the fault location in making relaying decisions, these estimates need only be accurate enough to determine whether the fault is in the appropriate zone of protection. More accurate estimates of the fault location are desirable for inspection, maintenance, and repair of the actual fault. For a long transmission line, for example, an error in fault location of a few miles may be acceptable for a relaying decision but would represent a long walk in rough terrain. While analog techniques for fault location have been reported,[21] the area of fault location represents another active area of algorithm development. A first step in producing a more accurate estimate of fault location from a relaying algorithm is to increase the data window significantly. As seen in Section 4.5, the accuracy of the estimate increases with the length of the window. A longer data window does not completely solve the problem in the presence of fault resistance, however.

The difficulty produced by fault resistance can be appreciated by examining Figure 9.15(a). The Figure shows a single-phase one-line diagram of a fault with resistance R_f at a distance k of the line with line impedance Z connected between sources with Thévenin impedances of Z_S and Z_R respectively. The relationship between the current I and voltage V measured by the relay is

$$V = kZI + I_fR_f \qquad (9.33)$$

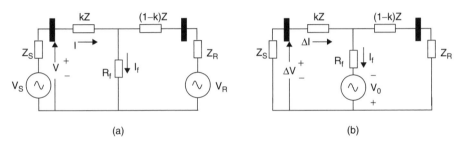

Figure 9.15 One line diagram of a faulted line. (a) The post-fault model. (b) The incremental model

where I_f is the current through the fault resistance. The current I_f includes contributions from both ends of the line and is unknown to the relay. If the current I_f is in phase with the current I, dividing (9.33) by I gives

$$\frac{V}{I} = kZ + \gamma R_f \qquad (9.34)$$

where $\gamma = I_f/I$ is real. The imaginary part of the computed impedance is correct and can be used for fault location.

The phase relationship will be true if the source at the remote end makes no contribution to I_f, or for specific values of V_R and Z_R. In general, however, γ is not real and Equation (9.34) has errors in both the real and imaginary parts. Some additional information about the system must be used to resolve the uncertainty. For example, information from the remote end could be used to determine I_f and Equation (9.33) could be multiplied by the conjugate of I_f to yield:

$$k = \frac{\text{Imag}\{V \times I_f^*\}}{\text{Imag}\{Z \times I \times I_f^*\}} \qquad (9.35)$$

since $I_f I_f^* R_f$ is real. The requirement of communication between the two ends is a limitation, and a number of techniques for accurate fault location using information from only one end of the line have been suggested.[21-25] All of these techniques take advantage of the additional information provided by the pre-fault currents and voltages seen by the relay.

If we let I_p and V_p denote the pre-fault value of current and voltage at the relay terminals in Figure 9.15(a), and define ΔI and ΔV as:

$$\Delta I = I - I_p$$
$$\Delta V = V - V_p$$

then the circuit in Figure 9.15(b) describes the relationship between the incremental quantities. The Δ quantities can be computed by subtracting the stored pre-fault values from the post-fault values. A similar diagram is obtained if the transmission line is described by the telegraph equation Equation (9.1).[21-24] The use of a distributed model of the line seems unnecessary for reasonable length lines. If the current ΔI is assumed to be in phase with I_f, Equation (9.36) can be multiplied by the conjugate of ΔI to obtain[21]

$$k = \frac{\text{Imag}\{V \times \Delta I^*\}}{\text{Imag}\{Z \times I \times \Delta I^*\}} \qquad (9.36)$$

Similar expressions are obtained for the distributed line model in a form that is appropriate for Fourier[23] or Laplace[24] transform methods. The solution of the nonlinear algebraic Equations for k, I_f, and R_f using the Newton-Raphson Method has also been proposed.[22] The use of a current distribution factor which relates the incremental positive sequence current to the fault current has also been suggested.[25]

9.8.2 Fault location based on traveling waves

When a fault occurs on a transmission line, it induces step wave changes in voltages and currents which propagate towards the ends of the transmission line as discussed in Sections 9.2 and 9.3. Consider a single phase transmission line shown in Figure 9.16(a). A fault at a distance x from one terminal will produce traveling waves which travel in both directions, arriving at instants T_1 and T_2 respectively at the two terminals. If the arrival of the waves is identified on a common time reference (in this case a GPS time signal), then knowing the difference between T_1 and T_2 and the velocity of propagation the distance to the fault can be estimated by

$$x = L/2 - v(T_1 - T_2) \qquad (9.37)$$

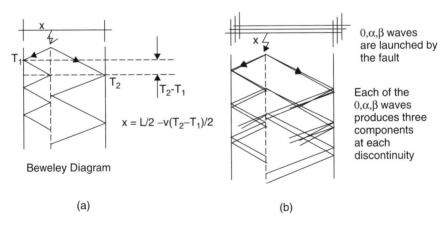

Figure 9.16 Traveling waves generated by a transmission line fault. (a) Waves on single-phase line and (b) Waves on three-phase line

The lattice diagram of wave propagation is commonly referred to as 'Bewley Diagram'. On a two conductor line the velocity of propagation is very nearly equal to the speed of light, and knowing the length of the line L, the fault location can be determined.

For a three-phase line, as discussed in Section 9.3, there are three modes of propagation launched by a fault: 0, α, and β. The latter two waves propagate at the speed of light, while the 0 wave propagates at a slower speed (approximately 70% the speed of light for most overhead lines). Each wave, upon reaching the line terminal, produces three modes of reflected waves, and the multiplicity of propagating waves soon make the wave arrival detection quite difficult. It is therefore necessary to determine the arrival of first of the α, and β waves in order to determine the fault location. It has been found in practice that current waves are easier to detect than the voltage waves when the voltage transducers may be capacitive voltage dividers.[27,28] Also, phase currents are converted to α, and β currents by using Equation (9.9) and discriminant functions (see Equation (9.19)) are used to isolate arriving waves from the terminal current signals. It has been reported in references 27 and 28 that the fault location performance of such devices are excellent. The reader should consult those references for additional information.

9.9   Other recent developments

Many recent developments have a direct bearing on computer relaying. We will consider some of these developments now.

The fiber optic communication networks have been mentioned several times. This technology permits very high data transmission rates (in several megabits per second) and can be used for protection as well as for other communication needs. Traditionally, protection tasks have demanded an independent communication channel. In the case of the fiber optic links, a separate fiber within a bundle dedicated to the relaying tasks may be an acceptable alternative. The most common fiber system at present is one where the ground wires on transmission lines are replaced by fiber-core ground wires. In time, it seems reasonable that such a fiber network will provide sufficient channel capacity, high redundancy, and economic incentive to become an integral part of protection systems.

Fiber optic systems are also being used to communicate relaying and control signals between the substation yard and control house where the relays are located. These links are not disrupted by electromagnetic interference, and with multiplexing could replace great deal of wiring from traditional substation designs. On the other hand, these systems require that electronic systems be placed within the yard, and hence that some enclosures and power supplies be distributed throughout the yard. In any case, as such systems become commonplace, computer relaying will become their natural extension.

Developments in electronic transducer technology were also mentioned in Chapter 1. A number of electronic, digital current transformers have been tested and are being tried on power systems throughout the world. A corresponding development in voltage transformers has not been as rapid, except the in case of gas-insulated substations where a controlled environment between ground and the high voltage conductors permits the installation of high precision electronic digital voltage transformers. These newer electronic voltage and current transducers, since they can provide sampled data directly, are a natural complement of a substation computer system. The reader is referred to relevant technical literature on the subject for additional details.[28-30]

9.10 Summary

In this chapter, we have introduced some concepts which are not in the main stream of computer relaying applications. Traveling wave relaying has held high promise, but has yet to be deployed in large quantities. Differential relaying of transmission lines based either on data samples or on phasors has become quite useful to utilities having sufficient communication capability and long transmission lines or transmission lines with multiple taps. Fault location technology for overhead lines has progressed sufficiently so that locating faults with distance estimation is quite accurate. Fault location based on wave propagation has also been accepted and is reported to provide very good results, although understandably the number of such installations is rather limited. Newer developments in technology arrive with astonishing speed, and it is to be expected that in the field of computer relaying we will see many such technologies adopted in the coming years.

Problems

9.1 Verify by direct substitution that expressions (9.2) are the solutions of Equation (9.1). Show that the characteristic impedance is $\sqrt{(L/C)}$ and the velocity of propagation is $1/\sqrt{(L/C)}$.

9.2 Given that the zero sequence inductance of a transmission line is three times the positive sequence inductance, and that the zero sequence velocity of propagation is 75% of the positive sequence velocity, what is the ratio of the zero sequence to the positive sequence characteristic impedance?

9.3 If a transmission line with a characteristic impedance of Z ohms terminates in a load impedance of Z', what is the reflection coefficient at this terminal? For a purely resistive Z, calculate the reflected wave if the incident wave is a step function and Z' is (a) purely inductive, and (b) purely capacitive.

9.4 Show that the Clarke and Karrenbauer transformation matrices given by Equations (9.9) and (9.18) are similarity transformations on the L, C, LC, and CL matrices of a transposed transmission line.

9.5 Validate the results given in Table 9.1. Often arguments of symmetry simplify the computations. Remember that, in each case, the fault is simulated by the application of negative prefault voltages at the point of fault.

9.6 Determine the relationship between the Karrenbauer components for each of the faults listed in Table 9.1.

9.7 An unsymmetrical fault will create mixed reflected waves when a pure modal wave strikes the fault. Determine the reflected waves of $0, \alpha, \beta$ modes when a pure α mode wave strikes a phase b-c fault.

9.8 Assume that a generator of infinite short-circuit capacity is connected to a transmission line with a characteristic impedance of Z, and a velocity of propagation v. Calculate and plot the discriminant functions given by Equation (9.20) through several successive reflections for a fault at a distance d. Thus show that the discriminant functions will maintain their relative values for substantial periods even for faults at short distances from the relay.

9.9 Assume that the capacitances of a three-phase transmission line can be neglected. For such a case, perform the synchronization of phasors at the two ends of the line. Assume zero mean random measurement noise of certain variance. Determine the uncertainty in synchronization when all phasors are measured with the assumed noise.

9.10 For the three-terminal line shown in Figure 9.15, assume a π-section representation for each of the line sections. If the voltages and currents at the terminals are available, determine a formula for calculating the charging current contribution from the capacitors at the tap point. This calculated compensation must be used to create an accurate differential current.

9.11 Show that the traveling wave differential currents given by Equations (9.31) and (9.32) become the phasor differential currents of Equation (9.27) when steady state conditions are reached on the line. Recall that the steady state condition can be viewed as a superposition of two constant waves traveling in opposite directions – the standing wave phenomenon.

References

[1] Rudenberg, R. (1968) *Electrical Shock Waves in Power Systems*, Harvard University Press, Cambridge, Massachusetts.
[2] Stevenson, Jr., William D. (1982) *Elements of Power System Analysis*, 4th edition, McGraw-Hill Inc.
[3] Sunde, Erling D. (1949) *Earth Conduction Effects in Transmission Systems*, D. Van Nostrand Company; (1968) Dover Publications, New York.

[4] Bewley, L.V. (1933) *Traveling Waves on Transmission Systems*, John Wiley & Sons, Inc. New York; (1963) Dover Publications, New York.

[5] McLaren, P.G. (1988) Traveling waves and ultra high speed relays, Chapter 6 in Microprocessor Relays and Protection Systems, IEEE Tutorial Course, IEEE Special Publication no. 88EH0269-1-PWR.

[6] Dommel H.W. and Michels, J.M. (1978) High speed relaying using traveling wave transient analysis, IEEE paper no. A78, pp. 214–219.

[7] Mansour M.M. and Swift, G.W. (1986) Design and testing of a multi-micro-processor traveling wave relay, IEEE Trans. on Power Delivery, vol. 1, no. 4, pp. 74–82.

[8] Vitins, M. (1978) A correlation method for transmission line protection, IEEE Trans. on PAS, vol. 97, no. 5, pp. 1607–1617.

[9] Kohlas, J. (1973) Estimation of fault location on power lines, Proceedings of the 3rd IFAC Symposium, The Hague/Delft, the Netherlands, pp. 393–402.

[10] Johns A.T. and Aggarwal, R.K. (1980) New ultra high speed directional blocking scheme for transmission line protection, *Developments in Power System Protection*, IEE Conference Publication no. 185, London, pp. 141–145.

[11] Hedman, D.E. (1965) Propagation on overhead transmission lines – I: Theory of modal analysis; II: Earth conduction effects and practical results, IEEE Trans. on PAS, vol. 84, pp. 200–211.

[12] Engler, F. Lanz, O.E. Hanggli, M. and Bacchini, G. (1985) Transient signals and their processing in an ultra high speed directional relay for EHV/UHV line protection, IEEE Trans. on PAS, vol. 104, no. 6, pp. 1463–1474.

[13] Chamia M. and Liberman, S. (1978) Ultra high speed relay for EHV/UHV transmission lines – development, design and application, IEEE Trans. on PAS, vol. PAS-97, no. 6, pp. 2104–2116.

[14] McLaren, P.G., Rajendra, S., Shahab-Eldin, S. and Crossley, P.A. (1985) Ultra high speed distance protection based on traveling waves, International Conference on Developments in Power System Protection, IEE Conference Publication no. 249, pp. 106–110.

[15] Takagi, T., Baba, J.I., Uemura, K. and Sakaguchi, T. (1978) Fault protection based on traveling wave theory, Part II: sensitivity analysis and laboratory tests, IEEE paper no. A 78, pp. 220–226.

[16] Thorp, J.S., Phadke, A.G., Horowitz, S.H. and Begovic, M.M. (1987) Some applications of phasor measurements to adaptive protection, Proceedings of the Fifteenth PICA Conference of IEEE, Montreal, pp. 467–474.

[17] Phadke A.G. and Hankun, H. (1986) Current differential relaying of multi-terminal lines with microprocessors, Proceedings of the Minnesota Power Systems Conference, Minneapolis, pp. 1–8.

[18] Takagi, T., Miki, T., Makino, J. and Matori, I.M. (1978) Feasibility study for a current differential carrier relay system based on traveling wave theory, IEEE paper no. A 78, pp. 132–133.

[19] Stranne, G., Kwong, W.S. and Lomas, T.H. (1986) A current differential relay for use with digital communication systems: Its design and field experience, 13th Annual Western Protective Relay Conference, Washington State University, Spokane, Washington.

[20] Souillard, M., Sarquiz, P. and Mouton, L. (1974) Development of measurement principles and of the technology of protection systems and fault location systems for three-phase transmission lines, CIGRÉ Paper no. 34-02.

[21] Takagi, T., Yamakoshi, Y., Yamaura, M. *et al.* (1982) Development of a new type fault locator using the one-terminal voltage and current data, IEEE Trans. on PAS, vol. PAS-101, no. 8, pp. 2892–2898.

[22] Westlin, S.E. and Bubenko, J.A. (1976) Newton-Raphson technique applied to the fault location problem, IEEE PES Summer Meeting, A76 334–3.

[23] Takagi, T., Yamakoshi, Y., Baba, J. *et al.* (1981) A new algorithm of an accurate fault location for EHV/UHV transmission lines: Part I – Fourier transformation method, IEEE Trans. on PAS, vol. PAS-100, no. 3, pp. 1316–1323.

[24] Takagi, T., Yamakoshi, Y., Baba, J. *et al.* (1982) A New algorithm of an accurate fault location for EHV/UHV transmission lines: Part II – Laplace transformation method, IEEE Trans. on PAS, vol. PAS-101, no. 3, pp. 564–573.

[25] Eriksson, L., Saha, M.M. and Rockefeller, G.D. (1985) An accurate fault locator with compensation for apparent reactance in the fault resistance resulting from remote-end infeed, IEEE Trans. on PAS, vol. PAS-104, no. 2, pp. 424–436.

[26] Gale, P.F. (1993) Overhead line fault location based on travelling waves and GPS, precise measurements in power systems conference, Arlington, Virginia.

[27] Gale, P.F., Taylor, P.V., Naidoo, P. *et al.* (2001) Travelling wave fault locator experience on ESKOM's transmission network, Developments in Power System Protection, IEE Conference Publication no. 479.

[28] Mouton, L., Stalewski A. and Bullo, P. (1978) Non conventional current and voltage transformers, *Electra*, no. 59, pp. 91–122.

[29] Hild, H.A., Stern, C., Gambale, J.C. and Sun, S.C. (1975) Field installation and test of an EHV current transducer, Trans. of IEEE on PAS, vol. 94, no. 1, pp. 37–44.

[30] Subjak, Jr., J.S. (1975) An EHV current transducer with feed-back controlled encoding, Trans. of IEEE on PAS, vol. 94, no. 6, pp. 2124–2130.

10

Wide area measurement applications

10.1   Introduction

The concept of adjusting the protection systems to adapt them to prevailing power system conditions was first embodied in 'Adaptive Relaying' which will be discussed more fully in the following section. In more recent years the technology of Wide Area Measurements based on synchronized Phasor Measurement Units (PMUs) has become a vehicle for gathering very precise information about the power system in real time.[1] These measurements are finding applications in several areas of power system operations including monitoring, control and protection. Phasor measurement technology is able to create a precise snapshot of the power system, which can be refreshed at rates approaching once per cycle of the power frequency. Being able to track the power system performance through normal and emergency conditions is clearly of great importance in adapting control and protection systems so that their response to system events is appropriate and optimal. Several promising applications of WAMS based adaptive relaying will be considered in Section 10.6.

10.2   Adaptive relaying

Adaptive relaying is a subject of relatively recent origin. It has been defined as follows.[2-6]

Adaptive Protection is a protection philosophy which permits and seeks to make adjustments to various protection functions in order to make them more attuned to prevailing power system conditions.

The key concept is to change something in a protection system in response to changes in the power system caused by changing loads, network switching operations, or faults. To a certain extent all existing protection systems must accommodate power system changes. Often this is achieved by making the relay settings correct for all conceivable network conditions. For example, zone-3 of a relay must be set such that it will cover the longest neighboring circuit (highest impedance) regardless of whether or not any in-feed from other lines is present. Considerations such as these often result in a protection system design that is deficient on two counts: firstly not all possible system contingencies can be anticipated during the design of the protection system, and secondly, the settings are not the best ones possible for any single system configuration. Nevertheless, to the extent that a relay setting or relay design caters for variable system conditions, it is an adaptive setting and in this sense present protection practices are adaptive to a certain extent.

Another point to be made is that the distinction between control and protection functions becomes blurred when one begins to consider adaptive protection functions. Many protection functions already encompass control: reclosing of circuit breakers is a case in point. As we begin to modify relay performance based upon changing conditions on the power system, we approach the classic concept of feed-back control. Indeed, adaptive relaying is in fact a feed-back control system.

Since adaptive relaying implies that relays must adapt to changing system conditions, a hierarchy of computer relays with communication links must be assumed to exist. The communications could be to other equipment within a substation, or to computer networks of remote substations. Consequently, one could visualize an adaptive protection system which adapts to its substation environment, or one which is responsive to system needs. Clearly the latter is more comprehensive, but also needs long distance communication channels. This becomes practical with the modern WAMS systems. Local communication within the substation is easy to achieve, and quite possibly adaptive features which depend upon local communication alone will be the first ones to be implemented.

Even where long distance communication is necessary, some adaptive features call for a great deal of real-time data to be obtained from remote locations, while others require modest amounts of non-real-time data. Adaptive systems requiring modest data transactions are sure to be implemented before those requiring large amounts of real-time data. At this time, fiber optic communication links offer the only feasible medium for such voluminous data transfers. Fiber optic links are just now beginning to be installed on power systems. As such installations become more commonplace, many more adaptive relaying concepts are likely to be accepted by the industry.

10.3   Examples of adaptive relaying

The reader is referred to the literature for a more complete account of adaptive relaying possibilities.[2−7] Many of the examples discussed in the literature are much better discussed in the context of WAMS which will be covered in Sections 10.5 and 10.6. In this section we will consider only those adaptive protection functions which require modest amounts of information external to the relay, which is often available locally. We will now present some examples to illustrate the principles involved.

10.3.1 Transmission line protection[8]

Traditional protection of multi-terminal lines calls for many compromises. Consider the three-terminal line shown in Figure 10.1. The zone settings of the relay at location A should be such that it remains secure whether or not the tap is in service. Consider a fault at F. The distance relay at bus A sees a fault impedance of

$$Z = Z_A + Z_F(I_A + I_C)I_A \qquad (10.1)$$

where I_A and I_C are the contributions to the fault from terminals A and C respectively. Equation (10.1) corresponds to a phase fault – similar equations hold for ground faults also. If one does not wish to over-reach the end B for all system conditions, the zone setting must be less than $Z_A + Z_B$ whether or not the tap at C is in service. If one sets zone-1 of relay A at 90% of $(Z_A + Z_B)$, it will protect a smaller portion of the distance from the tap point to the end B when the tap is in service. The zone-1 setting is thus less than desirable (i.e. 90% of the line) when the tap is in service.

One could restore zone-1 setting to its proper value under all conditions by making the relay adapt to system conditions. As mentioned previously, adaptability can be of different varieties depending upon the demands one can make on available communication channel capacity. For example, the status of breakers at B and C could be communicated to the relay at A. When breaker C status is known to the relay at A, zone-1 setting there can be 0.9 $(Z_A + Z_B)$ or 0.9 $[Z_A + Z_B(I_A + I_C)/I_A]$ depending upon whether the breaker at C is open or closed. To get this information from one terminal to the others, relatively small amounts of data need to be communicated between terminals, and the status data can be sent whenever the

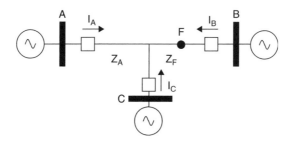

Figure 10.1 Three-terminal line with a fault at F

status changes – they need not be sent at real-time data rates. A reasonable value for the ratio of currents $(I_A + I_C)/I_A$ may be used in making the zone calculations. Similar considerations apply to relays at B and C.

The approximations introduced by the assumed values of currents in the procedure described above can be eliminated if the actual equivalents of the network at each terminal are sent to every other terminal of the multi-terminal line. This will require more data to be transmitted, but these too are not real-time data. Knowing the Thévénin equivalent for the other terminals as well as the status of the associated breakers, adaptive determination of the zone settings is possible. The procedure can be improved even more if the terminal which determines that the fault is between the tap point and the terminal communicates this information in real time to all other terminals. Since the segment of the line on which the fault has occurred is now known at each terminal (along with the equivalent circuit for the entire multi-terminal system), zone-1 can be set to reach the desired distance in the faulted segment from each terminal. The only margin which must be allowed is the usual one for any possible transient over-reach.

The Thévénin equivalent circuits require significantly greater amounts of data to be communicated to each terminal. Sequence impedance matrices and voltage vectors must be sent from a central location where the equivalent circuits can be determined. Although these are not real-time data in the sense that they must be sent while the fault is on the system, they must be computed and communicated often enough to keep the data current. The most accurate multi-terminal line protection can be obtained if voltage and current measurements (either sampled data or phasors) are communicated from each terminal to all the others in real time. Indeed, in this case the differential protection principle described in Section 9.7 should be used, instead of the stepped distance protection.

10.3.2   Transformer protection

Transformer protection using percentage differential relays was discussed in Section 2.4. It was pointed out there that the slope of the differential characteristic is adjusted to over-ride the false differential currents produced due to mismatches in current transformer ratios, changing taps in a tap changing transformer, and unequal errors in current transformers in the primary and secondary windings of the transformer. Typical percentage slopes of the differential relays are set at 40% to accommodate these false differential currents. Two of these three effects, viz. current transformer ratio mismatch and changing tap position could be taken into account by an adaptive percentage differential relay.

Consider the adaptive differential relay shown in Figure 10.2(a). The relay can monitor input primary and secondary currents i_1 and i_2 when the transformer is carrying load current. Under healthy conditions, these two currents should be equal. If they are not, there will be a false differential current which is due to the three effects mentioned above. In this case, the relay will estimate a multiplication

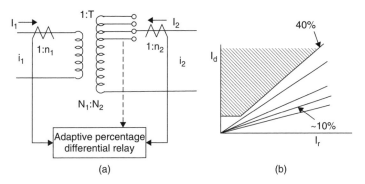

Figure 10.2 Adaptive percentage differential relay. (a) Relay with main transformer tap input to the relay. (b) Reduction in the percentage differential slope

factor k_2 to be applied to the secondary current i_2 so that the two currents are equal:

$$i_1 = k_2 i_2 \qquad (10.2)$$

The factor k_2 is an estimate of the CT ratio mismatch and the effect of the tap changer position on the main power transformer. (The tap changer position of the main power transformer can also be sensed directly by the differential relay if such a sensor indication is available.) This factor will be used in calculating the differential and restraint currents in the differential relay algorithm, so that the only remaining cause of the false differential current will be the effect of unequal CT errors during fault conditions. Having removed two of the three causes of false differential current, the percentage differential slope of the relay can now be adjusted downward as illustrated in Figure 10.2(b).

Clearly the reduced percentage differential slope increases the sensitivity of the relay, and thus it is made capable of sensing lower grade faults in the transformer.

10.3.3 Reclosing

In many countries (including North America) the standard practice for handling phase-to-phase and phase-to-ground faults is to used high speed three-phase clearing, followed by high speed reclosing, and on detecting a sustained fault to follow up with some number of automatic reclosing, and a final lock-out if the fault is permanent. The first high speed reclosing is without supervisory interlocks (i.e. without checking for voltages on either side of the breaker), and the speed of high speed reclosing is determined by the system voltage – varying between 20 and 40 cycles. The automatic reclosing functions are usually controlled by conditions on the network, and usually employ some checks and interlocks.

Since the high speed reclosing function does not check system conditions, it is possible to reclose on an existing fault in the case of a permanent fault. Generators which are near the fault are thus likely to be subjected to a repeat stress induced by closing into a fault. In particular, the closing into a phase-to-phase fault is likely to cause a severe shock to the generator and a consequent harmful effect on the machine shaft life. It is therefore desirable to use an adaptive high speed reclosing scheme which tests the system voltages before attempting the high speed recluse, and blocks it if an existing fault is discovered.[3] This is particularly desirable when operation of relays near a generator is under consideration.

A simplified system with a generator, a transmission line, and a circuit breaker is illustrated in Figure 10.3. It is assumed that the circuit breaker is capable of operating each pole individually. At high speed reclose time, one pole of the circuit breaker is closed. The choice of which pole to close depends upon which phase had the original fault. For example, if the initial fault was on phase b, the reclose operation is initiated on phase a or c. Let us assume that phase a is selected for reclosing first. The voltages on all three phases are next measured (usually within one period of the fundamental frequency), and depending upon the voltages on the three phases the condition of the line can be established. Following possibilities exist:

(a) There is no fault on the line. In this case phase-a voltage will be normal, and phases b and c will have induced voltages which are in phase with phase-a voltage, and their magnitudes will depend upon the conductor placement on transmission line towers. These being known, could be pre-set in the relay as norms for induced voltages for that line configuration. In this case the high speed reclose operation could be completed on the remaining phases.

(b) There is a phase-a-to-ground fault on the line. In this case all phase voltages will be low, and phase a current will be high.

(c) There is a phase-b-to-ground fault on the line. In this case phase b voltage will be low, and phase c will have an induced voltage.

(d) There is a phase-b-to-phase-c fault on the line. In this case the induced voltages on phases b and c will be exactly equal to each other.

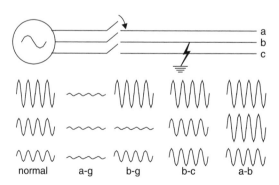

Figure 10.3 Reclosing on a transmission line with a phase-b to ground fault. The circuit breaker tests the line with a single phase reclose and blocks further high speed reclose operations if a standing fault is detected

(e) There is a phase-a-to-phase-b fault on the line. In this case the voltage on phase b will be equal to the voltage on phase a, and c phase will have an induced voltage.

In the last four cases further reclosing operation will be blocked, and the first (closed) pole opened again. This adaptive reclosing scheme prevents the generator to be subjected to a multi-phase fault on high speed reclosing, which is the most damaging as far as generator shaft life is concerned.

10.4 Wide area measurement systems (WAMS)

Wide area measurements through synchronized phasor measurements have become commonplace on many power systems around the world. This technology offers precise measurements of positive sequence voltages and currents at remote locations.[1] The instruments which perform these measurements are known as Phasor Measurement Units (PMUs). A generic PMU is shown in Figure 10.4. The basic structure of a PMU is very much like that of a computer relay. Input signals (consisting of voltages at buses and feeder currents) are sampled and the sampled data converted to positive sequence quantities as per the technique given in Chapter 5. The measurements are time-tagged using a GPS clock. The measurements can be refreshed as frequently as once every cycle. The details will be found in reference [1] cited above.

The outputs of the PMUs with the accompanying time tags are available for communication to local or remote sites. Where remote communication with several PMUs is involved, the communication is through hubs which are known as Phasor Data Concentrators (PDCs). PMUs and PDCs form a hierarchical structure discussed in the next section. Main features of PMUs and PDCs are defined in an IEEE standard (C37.118 and its succeeding revisions) so that equipment manufactured by different manufacturers can be inter-operable.

10.5 WAMS architecture

The WAMS system consists of PMUs where the measurements are carried out, and PDCs where inputs from several PMUs are collected and passed on to various applications as needed. This hierarchical arrangement is illustrated in Figure 10.5.

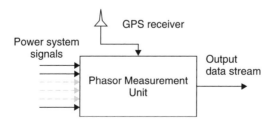

Figure 10.4 A generic phasor measurement unit. Three-phase voltage and current waveforms are sampled and then converted to positive sequence measurements

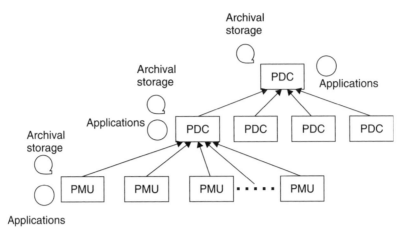

Figure 10.5 Hierarchy of PMUs and PDCs to manage the wide area measurement system

The task of the Phasor Data Concentrator can be briefly summarized as follows. It matches the time tags of data received from various PMUs so that a synchronized data stream is created for applications, and communicated to upper levels of hierarchy for further data concentration. PDCs (as well as PMUs) store archival data which can be used for post-mortem analyses of major system events. In addition the PDCs may also have a GPS receiver so that latency of data received from various PMUs can be measured, and outliers in propagation delays flagged for corrective action to be taken. Applications of phasor data may reside at PMU levels as well as at PDC levels. Clearly at higher levels of the hierarchy the volume of data collected increases, and so does the latency of the data. There is thus a natural selection in applications which can be implemented at various levels of the hierarchy. At PMU level, relatively fast actions can be taken using locally available data. At regional or central control centers, applications requiring data from wider areas with longer delays are more appropriate. This distinction in applications holds for system monitoring and control applications, as well as for the protection applications using WAMS.

The PMU and PDC architecture also permits data transmission in opposite direction on every communication link. Usually the data in the downward direction is sparse and infrequent, mostly dealing with system management functions.

10.6   WAMS based protection concepts

Many of the ideas of adaptive protection become particularly attractive when wide area measurements are used to achieve adaptability. Because of the delays in gathering data at PDCs (typically of the order of 30–100 milliseconds), the adaptive protection functions which are based on WAMS must be relatively slower acting. Most backup and system protection functions are of this type, and are ideal for adaptability using WAMS data. It should also be stated that as of now these concepts are proposals for research and development for practical implementations to follow in the future. It is also likely that newer ideas on such protection applications will be forthcoming in the future, and the reader is advised to consult technical publications for evolving ideas.

We will now examine some of these concepts which have been discussed in recent publications.[9,10]

10.6.1 Adaptive dependability and security

During normal operation of power systems, a failure to trip quickly when a fault occurs can lead to system instability. It is therefore imperative that a fault is cleared in primary clearing time without fail. This is the requirement of high dependability in protection during normal system operation. This bias is appropriate when the power system is in normal state. The bias towards dependability invariably produces a reduction in security of protection.[11] (Also see Section 2.1) However, when the power system is in a stressed state a false trip – a consequence of reduced security – is far more destructive to the power system, as it may lead to cascading failures and a wide-spread blackout. WAMS data could be used at the control center, and an assessment made of the state of the power system. When the system is determined to be in a stressed state, it would be desirable to alter the protection system bias in favor of increased security, with a possible reduction in dependability. In doing so, it is accepted that there may be a fault for which there may be a slight increase in the probability that the fault will not be cleared in normal time, but the chance of a false trip is significantly reduced.

Figure 10.6 illustrates the principle of this scheme. From prior off-line analysis of the power system, critical locations of protection systems are determined so that a false trip of relays at those locations will increase the possibility of cascading failures, and adaptive supervision of the system is desirable. Consider the normal complement of protections at such a location. There are three independent protection systems, and in order to achieve high dependability during normal system operating conditions the outputs of the three protection systems are arranged in a logical 'OR' configuration. This means that operation by any one of those protection systems will lead to the trip of the circuit breaker. Of course, insecure operation of any one protection system may lead to the start of cascading failures.

Now consider the scenario at the control center where the WAMS based estimation leads to the conclusion that the power system is in a critical state. Under these conditions, the protection system bias could be changed in favor of increased security. This is achieved by modifying the logic of the critical protection system to a 'VOTE', so that at least two of the protection systems must agree that the fault requires tripping of the circuit breakers before the trip is issued. This is best achieved when the protection system consists of computer relays, and the logic is

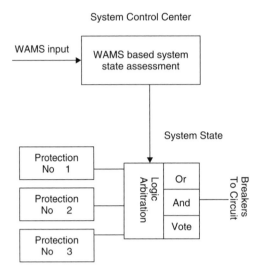

Figure 10.6 Adaptive control of dependability and security of protection systems based on WAMS

not realized by hard-wired relaying systems. Of course this principle can be applied at more locations if they are determined to be critical.

It should also be noted that there is always a possibility that two out of three protections may mis-operate, leading to an insecure operation. However, by changing the logic to 'VOTE', the level of security has been increased very significantly.

10.6.2 *Monitoring approach of apparent impedances towards relay characteristics*

Several analyses of blackouts have revealed that some protection settings were made with certain assumptions about the power system – for example its expected load level or stability oscillations following a major event. The relay settings made are appropriate for the assumed system condition. However, over the years the power system changes in its structure, operating strategy and loading patterns. Under ideal circumstances the relay settings made would be revised with each significant change in the power system. In most power systems this is not done on a regular basis, because of shortage of manpower or scheduling difficulties. In any case, it has been observed that as years go by, the prevailing relay settings are not appropriate for the existing power system state and the criticality of some of the relay settings is not noticed by the protection engineers.

In the case of various distance relays, it is possible to monitor the apparent impedance trajectories during system load changes or stability swings with PMU measurements. Remember that the conditions of concern are balanced conditions, and the impedance relays would see the positive sequence impedance which can be obtained by dividing voltage phasors by feeder current phasors at the relay location in question. During load changes or stability swings it would be possible to compare the apparent impedance trajectory in relation to the relay characteristic, and depending on how close the apparent impedance comes to the relay characteristic, an alarm is communicated to the engineering office indicating that the relay setting may need to be reconsidered. This alarm is not meant to take action in real time, but to create a message to the engineering office that the relay setting may need a revision. This function would identify critical protection functions which may be in danger of causing a cascading failure due to unexpected operation of the relay. Of course similar procedure may be implemented for overcurrent relays as well.

Consider the power system shown in Figure 10.7. At a critical location, the relay characteristic would be available to the local PDC, and the PDC application would track the apparent impedance trajectory for load changes, as in Figure 10.7(a), or for stability swings, as in Figure 10.7(b). The margin settings used for creating the alarm would be set by protection engineers and would also be available to the local PDC application. This is one of the simplest applications of the PMU data, and since all signals and controls are local it is relatively easy to implement.

10.6.3 WAMS based out-of-step relaying

Out-of-step relays are commonly employed in systems where instability between two or more regions leads to a loss of synchronization and separation into two or more islands. Consider the system depicted in Figure 10.8. During transient stability swings of such a system, most transmission line relays will see an apparent

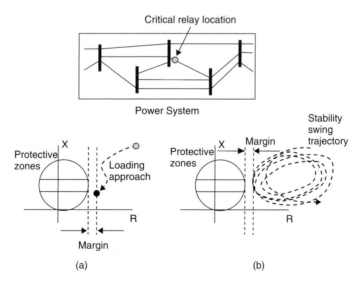

Figure 10.7 Detection of apparent impedance trajectory approaching zones of protection

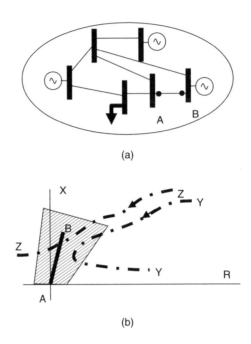

(a)

(b)

Figure 10.8 Stability swing excursions in the zone of protection of a relay

impedance trajectory in the R-X plane. For example, the distance relay at bus A for line A-B may see impedance trajectory Y-Y or Z-Z during two system swings of different severity. The first trajectory corresponds to a stable swing, while the second corresponds to an unstable swing. In any case, these impedance trajectories may enter one of the trip zones of the relay as shown in Figure 10.8(b). The out-of-step relays must be able to distinguish between a movement of the trajectory inside a zone of protection due to a fault, from one caused by the stability oscillations. Traditionally, this is achieved by using the criterion that, in case of a fault, the trajectory is traversed very quickly, while a stability oscillation creates a slow movement in the R-X plane.

Upon detection of an out-of-step condition, the next task of the protection system is to permit selective tripping for clearly unstable cases so that the power system is separated in islands, with a reasonable match between load and generation within each island. At present, the places where tripping is permitted are pre-determined, based upon simulations performed during system planning studies. However, a more appropriate procedure would be to determine both the nature of an in-progress swing as well as desirable locations for separation in real time.

The first part of this problem – determining whether a given transient swing which is in progress is one from which the system will or will not recover – is a difficult one, but can be expected to be solved by several of the techniques currently under investigation (such as parallel processing, use of transient energy functions etc.). For systems which behave like a two-machine system, the problem can be

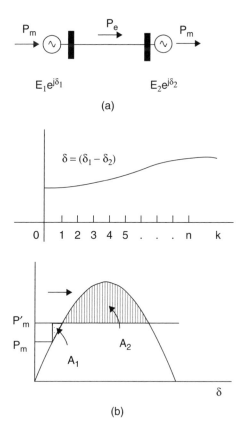

Figure 10.9   Prediction of stability swing

solved.[12] Assume that the two-machine equivalent of the system is known through dynamic state estimation. For a two-machine system, the stability oscillation can be predicted by the well known equal area criterion.

Consider the two-machine system shown in Figure 10.9(a) (representing a two area system equivalent) in which a loss of generation in one area has caused the steady state mechanical power to shift from P_m to P'_m leaving the transfer impedance unaffected. The swing equation of the two-machine system is

$$M\ddot{\delta} = P_m - P_e = P_{e\,max} \sin \delta \qquad (10.3)$$

where $(\delta = \delta_1 - \delta_2)$ and M is the equivalent moment of inertia of the two-machine system. From successive real-time measurements of voltage and current phasors at the terminals of the equivalent machines, a set of electrical power and corresponding rotor angle differences is measured in real time. Let P_k and δ_k {$k = 0,1,\ldots,n$} represent these measurements at instants $k\Delta t$, where Δt is the measurement interval. If we assume that the P–δ relationship is piece-wise linear, we may estimate P'_m

from the set of measurements (P_k, δ_k).[12] If $\hat{P}'_m$ is the estimate of P'_m

$$\hat{P}'_m = \frac{2}{(\Delta t)^2 (v^t v)} [v^t \delta - v^t 1 \delta_0] + \frac{1}{v^t v} \left[v^t BP + \frac{1}{3} v^t \overline{P} \right] \tag{10.4}$$

where P and δ are vectors of electrical power and angle measurements

$$\mathbf{P} = (P_0, P, \ldots \ldots P_{n-1})^t, \, \boldsymbol{\delta} = (\delta_1, \delta_2, \delta_n)^t$$

$\overline{P}$ is a vector of differences between every power measurement and the first power measurement P_0

$$\overline{\mathbf{P}} = [(P_1 - P_0), (P_2 - P_0), \ldots (P_n - P_0)]^t$$

and **v, B**, and **1** are matrices defined by

$$\mathbf{v} = (1, 4, 9, \ldots n^2)^t$$

$$\mathbf{B} = \begin{bmatrix} 1\,0\,0\,0...0 \\ 2\,2\,0\,0...0 \\ 3\,4\,2\,0...0 \\ .\,.\,.\,.\,.\,... \\ n\,.\,.\,.\,...2 \end{bmatrix}$$

$$\mathbf{1} = (1, 1, 1, \ldots 1)^t$$

Equation (10.4) is obtained by integrating Equation (10.3) over each sampling interval, and assuming that the electrical power changes linearly between the two measurements. The details of this derivation are left as an exercise (see Problem 2).

The new mechanical power P'_m having been estimated from the measurements of P and δ, the complete P–δ curve can be determined, and finally an equal area comparison made between area A_1 and the maximum available margin A_2 in Figure 10.9. If A_1 is smaller than A_2, stability is predicted. Otherwise, the swing will be unstable. In the simulations reported in the reference cited,[12] a reliable estimate of P'_m (and thus the inference about the nature of the stability oscillation) could be reached in about one-quarter of the period of the electromechanical oscillation.

A research project applied this theory to the interface between the states of Florida and Georgia in the US. Although there are a large number of machines in both of these systems, because of the peninsular nature of the Florida power system the behavior of the system during stability swings can be approximated by a two-machine equivalent – one representing the aggregate of Florida generators and the other the aggregate of the Eastern United States. This latter is of course a very large system, and approaches an infinite bus in relation to the power

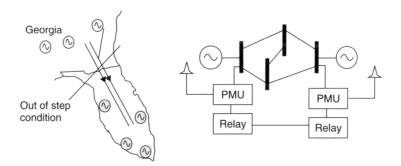

Figure 10.10   Adaptive out-of-step relaying for a two-machine like system

system of Florida. Thus, this (and similarly situated systems) are ideal candidates for application of the two-machine out-of-step relaying concepts developed above.

Figure 10.10 shows the system in question. PMUs collected phase angles from the two regions, and exchanged them with each other in order to construct a two-machine model of the stability event. In such a system, equal area criterion can be used to determine if an evolving swing is going to be stable or unstable.

The problem becomes much more complex on large integrated networks common in most modern economies. A possible solution for the stability problem in large networks with the use of real-time data from WAMS may benefit from a two-step approach. In the first step, the evolving swings of all generators of significant size may be observed, and over a reasonable observation window groups of coherent machines can be identified. See Figure 10.11(a). It is likely that in most practical cases coherency of generators can be established in about 200–400 milliseconds after the start of the disturbance.

Determination of coherency between machines facilitates the formation of coherently swinging groups as illustrated in Figure 10.11(a). When coherency groups have been determined there are two possible approaches to determining the outcome of the swings as illustrated in Figure 10.11(b). The first approach is to assume that the Extended Equal Area Criterion[13] can be applied, and the stability outcome determined as in the case of a two-machine stability problem. Briefly, the principle of Extended Equal Area postulates that at the time of a system split, each event of separation is similar to the behavior of a two-machine system. It is likely that as the system event progresses, other coherent groups may be formed after each split, and the Extended Equal Area principle can be applied sequentially to these evolving groups.

As an alternative to the Extended Equal Area approach, one could consider the swings of the coherent groups as time-series functions,[14,15] and a prediction of the time-series leads to either a zero-crossing of the relative angle which leads to stability, or to a monotonically increasing behavior which would indicate instability. Based upon this prediction of the outcome, one would determine appropriate control actions to be taken.

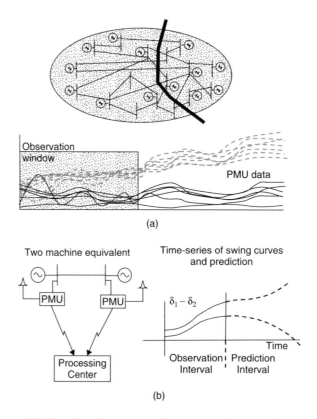

Observation window

PMU data

(a)

Two machine equivalent

PMU   PMU

Processing
Center

Time-series of swing curves
and prediction

$\delta_1 - \delta_2$

Time

Observation   Prediction
Interval   Interval

(b)

Figure 10.11   Out-of-step detection in large integrated networks

This subject remains an active area of research and newer developments can be expected to offer other possibilities for real-time stability determination.

10.6.4   Supervision of backup zones

Over-reaching zones of distance relays cover a large area in the R-X plane. This exposes them to inadvertent tripping when no fault is present on the system. In particular, zone-3 trips due to load excursions or due to stability swings are known contributors to loss of security in the protection system. Various schemes to restrict the exposure to zones of protection to such phenomena – for example using lens shaped or figure-8 characteristics – does reduce the likelihood of unwanted trips due to these effects. However, the fact remains that load power-factors can become very unusual during major disturbances due to unusual flows of active and reactive powers. As these phenomena are likely to occur when the power system is stressed and in danger of cascading into a blackout, it is appropriate to consider remedial actions which can be implemented using wide area measurements.

Zones 2 and 3 of step distance relays are both over-reaching zones. Zone-3 is traditionally recognized as one of the most difficult settings to be made because of the

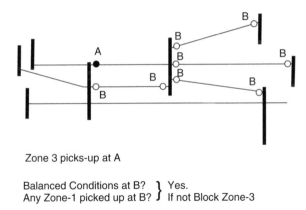

Zone 3 picks-up at A

Balanced Conditions at B? } Yes.
Any Zone-1 picked up at B? } If not Block Zone-3

Figure 10.12   Supervision of over-reaching zones of distance relays

need for adequate security. On the other hand, there are scenarios of contingencies when zone-3 is deemed essential for dependable clearing operation following a fault.[16] It is then necessary to investigate ways in which the over-reaching zones could be supervised by techniques which use wide area measurements.

PMUs located at neighboring buses offer an excellent tool for supervising the operation of over-reaching zones in order to avoid inappropriate operation during conditions of system stress.[10] Consider the relays located at buses identified by 'B' in Figure 10.12. If an over-reaching zone of relay at 'A' picks up, then PMUs at all the 'B' locations can be queried to determine if the apparent impedance seen by them represents a fault in the direct trip zones of any of those relays. If none of them see such a fault, then clearly the pick-up of an over-reaching zone at 'A' is due to something other than a fault, and it should be blocked. Note that this action is to be taken only for balanced loading conditions at A and B locations, since load encroachment or stability swings imply balanced operating conditions. The distances from which measurements are to be obtained are modest, so that communication delays should be insignificant.

10.6.5   Intelligent load shedding

Under-frequency load shedding and restoration is used in most power systems. Occasionally, load shedding under supervisory control has also been exercised as a part of a Remedial Action Scheme. It is clear that under-frequency load shedding does not operate unless the power system breaks into islands creating load generation imbalances, and there is an overall decay of system frequency in generation deficient islands. In many cases this action is too late, particularly if the islands are created in an unexpected configuration. It would be desirable to activate load shedding under supervisory control when significant loss of generation or load takes place, although no islands are formed and the frequency does not begin its decline.

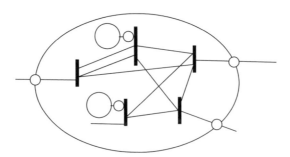

Figure 10.13   Real-time ACE determination from tie flow deviation to perform intelligent load shedding

A possible approach to determining loss of generation in a network is to determine deviation of the tie power flows (ΔT) from their scheduled values (see Figure 10.13). This is traditionally measured as an element of the Area Control Error (ACE = $\Delta T - B\Delta f$).[17] The other element of ACE is Δf, the deviation in system frequency, and B the average system load-frequency characteristic droop. Of course until the tie lines are opened, the frequency deviation is zero, and tie flow deviation alone signifies the generation shortfall. The shortfall can be used to estimate the amount of load to be shed. This latter need not match exactly the tie-flow deviation. Other factors to consider in such a scheme are the inertial response of generators to the initial disturbance. Ultimately, the goal of such a scheme is to make appropriate adjustments to system load in order to avoid more serious consequences in the future – such as a system-wide blackout.

10.6.6   Adaptive loss-of-field

A loss-of-field condition for a generator is detected by a distance relay connected at the terminals of a generator. If the generator field current is reduced due to a hardware failure or due to some inappropriate control setting in the excitation system of the generator, the generator may face one of two contingencies: excessive end-iron heating or steady state instability.[11] Both conditions can be detected by a distance relay with circular characteristics, as illustrated in Figure 10.14. When the limiting contingency is end-iron heating, the relay setting is based only upon the generator capability. However, when the limiting contingency is the approach of steady state instability, the relay characteristic depends upon the impedance of the generator and the Thévenin impedance of the power system, as seen from the machine terminals.

The distance relay setting consists of two concentric circles which are offset appropriately below the real axis, as shown in Figure 10.14. The inner circle is the actual stability limit, which, if breached, will lead to loss-of-synchronization of the generator and pole slipping and its eventual tripping. The outer circle is used to create an alarm for system operators if the apparent impedance seen by the relay

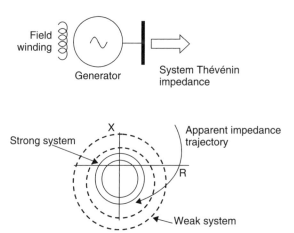

Figure 10.14   Adaptive loss-of-field relay. The relay settings are adaptively adjusted to the prevailing power system Thévénin impedance

continues its movement from outside the outer zone, through it, and approaching the inner zone which would create excessive end-iron heating or steady state instability.

The offset of the circular characteristics as well as their radii are dependent upon the two impedances mentioned above. Clearly, the machine impedance is a constant. However, the system Thévénin impedance will depend upon the strength of the power system. When the relay is set initially, it assumes certain conditions to exist on the network. However, as a catastrophic event unfolds, many important transmission elements may have tripped. This would create a weaker power system, and a larger Thévénin impedance. Under certain conditions, it is entirely possible that the weakened power system will lead to instability even before the outer zone of the relay (as set originally) is reached.

The state of the power system is continuously monitored at the control center through monitoring and state estimation tasks. It is then possible to determine in real time the changing system Thévénin impedance and communicate this information to the generator loss-of-field relay. This would make the relay setting reflect the prevailing power system state, thus improving security of the power system.

10.6.7   Intelligent islanding

Islanding of the power system is a natural extension of the out-of-step detection function. For disturbances which are going to lead to instability between two coherent groups of generators, it is a logical step to separate the power system into islands which contain each coherent group of generators and some load. It is clear that the loads or generators in the islands would have to be adjusted in order to continue operation at a safe system frequency. In islands with excess generation it will be necessary to shed generation of appropriate size, while for islands where there is excess load it will be necessary that load be shed. The loads and generators in the islands as initially formed would have to be in approximate match so that very large adjustments would not be necessary; also, both loads and generators should be available for control.

Wide area measurements available at the system control center will help identify desired boundaries of separation depending upon which generators have formed a coherent group, and what loads and generation within the islands to be formed would be controllable to the desired extent. Usually this requires out-of-step blocking at undesirable points of separation, and out-of-step tripping at desired locations for forming the islands. It may well be that some islanding scenarios are pre-calculated, based upon simulation studies and made available for action when system state assessment indicates that such control actions are called for.

10.6.8   System wide integration of SIPS

SIPS – System Integrity Protection Scheme is the acronym currently recommended by the IEEE Power System Relaying Committee to describe complex protection and control systems commonly known as Remedial Action Schemes (RAS) or Special Protection Schemes (SPS). The SIPS schemes usually bring in wide area measurements such as power flows, switch and circuit breaker status etc. and use pre-calculated responses to take action when power system reaches a state where such a control action is needed. Usually these schemes are designed to enhance power transfer capabilities between regions of the power system, and take immediate corrective action if a contingency condition occurs which does not permit that level of power transfer. The power transfer limit may be imposed by thermal loading limits, operating voltage limits, or stability limits of the network.

It is known that in many power systems several SIPS systems have been installed to take corrective action in response to many different sets of contingency conditions. As each SIPS is designed to remedy one particular operational constraint, it is possible that in an arbitrary system state, two or more SIPS may have conflicting effects on the power system, negating the effect of the cure provided by one or more of the installed SIPS.

A possible use of wide area measurement systems is to gather system-wide real-time data at the control center, and then determine which of the SIPS should be allowed to operate to steer the power system to a secure state or if a new control action is needed. The role of the wide area measurement system is to provide a sound basis to the system control center to use the most effective control strategy in any given situation.

10.6.9   Load shedding and restoration

The load shedding function is designed to maintain a balance between load and generation within a system. If the system has been islanded following a system disturbance of the type described above, its frequency will begin to move towards a new value determined by the mismatch between load and generation within the island. If the load is in excess of generation, the frequency will decay and it will become necessary to remove load soon enough before the frequency decays to such an extent that generators must be taken out of service in order to prevent generator damage. Usually load shedding is initiated by under-frequency relays installed in substations, and load is shed in several frequency steps in order to make sure that unnecessary load is not tripped beyond what is needed to arrest the frequency decline above the dangerous level. Load restoration is the reverse process: load is restored as the frequency returns to its normal value, and it is restored in steps with sufficient coordination delays in order to avoid any cycling effects between load shedding and restoration relays.

If the system does not form an island, the frequency remains substantially constant. In this case, the load shedding relays cannot act, as they do not see the necessary decay in frequency. However, the need to initiate load shedding is still there: each system should attain load-generation balance within its boundary to avoid overloading the tie-lines with its neighbors and risk a separation. This has been discussed in Section 10.6.5.

Recently an interesting approach to power system restoration following a blackout has been proposed.[18] The restoration scheme uses Artificial Neural Networks (ANNs) to facilitate service restoration in each island formed following a blackout. Within each island the first ANN determines load forecast based upon pre-blackout load data and cold-load pick-up for the types of load in the island. The second ANN determines the final expected island configuration taking into account the availability of various resources and the load that could be served upon restoration. Based upon these data, a switching sequence is generated which systematically restores lines and transformers with due attention being paid to loading and voltage constraints that may be exist. Wide area measurements were used to provide inputs to the ANNs, as well as to help determine the feasible switching sequence. The technique was demonstrated on a 162 bus, 17 generator test system.

Several other contributions to the subject of power system restoration can be found in the literature.[19] Fast and successful service restoration on a power system following a blackout is one of the most important counter-measures against catastrophic failures of power systems. As wide area measurements become integrated in power system management, it is to be expected that intelligent service restoration will receive increasing attention.

10.7   Summary

The synchronized phasor measurement technology was born out of the developments in computer relaying. The PMUs have been recognized as a major innovation in precise measurements on power systems, and several applications of these measurements to power system engineering are currently under development. In this chapter we have provided an overview of current thinking on using these measurements to improve power system protection – particularly as related to the problem of power system blackouts. It is expected that coming years will see implementation of these ideas. It is also the hope of the authors that future researchers will be motivated to more fully integrate the computer relays with wide area measurements in order to make the protection systems more dependable and secure under a variety of system operating conditions.

Problems

10.1 Assume that the *Thévenin* equivalents are known for the sources behind the terminals of a three-terminal transmission line shown in Figure 10.1. Assuming a fault occurs on one of the taps, and this fact is known to every other terminal, devise an algorithm which will determine the fault location accurately from each terminal. Also show that the effect of the fault arc resistance upon the distance estimate can now be eliminated.

10.2 Verify that the result given in expression (10.5) is the least-squares solution for estimating the new mechanical power transfer. The electrical power is assumed to be piece-wise linear between two measurements, and the rotor velocity at the end of one interval is the velocity at the beginning of the next interval.

10.3 Derive a result similar to Equation (10.5) if the rotor angle movements are calculated by the formula given in Chapter 14 of Reference 8 (below).

References

[1] Phadke, A. G. and Thorp, J. S. (2008) *Synchronized Phasor Measurements and Their Applications*, Springer.
[2] Phadke, A. G. Horowitz, S. H. McCabe, A. G. (1990) Adaptive automatic reclosing, Paper no. 34-204, CIGRÉ 1990 Session, August 26-September 1, 1990, Paris, France.
[3] Horowitz, S. H., Phadke, A. G., and Thorp, J. S. (1987) Adaptive transmission system relaying, Paper no. 87 SM 625-77, IEEE PES Summer Meeting, San Francisco, July 1987.
[4] Phadke, A. G., Thorp, J. S. and Horowitz, S. H. (1987) Impact of adaptive protection on power system control, Proceedings of the 9th PSCC Conference, Lisbon, pp. 283–290.
[5] Thorp, J. S., Horowitz, S. H. and Phadke, A. G. (1988) The application of an adaptive technology to power system protection and control, CIGRÉ, Paris.
[6] Phadke, A. G., Thorp, J. S. and Horowitz, S. H. (1988) Study of Adaptive Transmission System Protection and Control. Final report prepared for Oak Ridge National Laboratory by Virginia Polytechnic Institute.
[7] Rockefeller, G. D., Wagner, C. L., Linders, J. R., *et al*. (1987) Adaptive transmission relaying concepts for improved performance, Paper no. 87 SM 632-3, IEEE PES Summer Meeting, San Francisco.

[8] Stevenson, Jr., William D. (1982) *Elements of Power System Analysis*, Fourth edition, McGraw-Hill Inc.

[9] Phadke, A. G., Novosel, D. and Horowitz, S. H. (2005) Wide area measurement applications in functionally integrated protection systems, CIGRÉ Study Committee B5, Colloquium, October 2007, Madrid, Spain.

[10] Phadke A. G. and Novosel, D. (2008) Wide area measurements for improved protection systems", 8th Symposium on Power System Management, Cavtat, Croatia.

[11] Horowitz, S. H. and Phadke, A. G. (2008) *Power System Relaying*, Third Edition, John Wiley and Sons, Ltd.

[12] Centeno, V., De La Ree, J., Phadke, A. G. *et al*. (1993) Adaptive out-of-step relaying using phasor measurement techniques, IEEE Computer Applications in Power, vol. 6, no. 4.

[13] Xue, Y., Wehenkel, L., Belhomme, R. *et al*. Extended equal area criterion revisited [EHV power systems], IEEE Trans. on Power Systems, vol. 7, issue 3, pp. 1012–1022.

[14] Haque, M. H., Rahim, A. H. M. A. (1988) An efficient method of identifying coherent generators using Taylor series expansion, IEEE Trans. on Power Systems, vol. 3, pp. 1112–1118.

[15] Haque, M. H. (1996) Novel method of finding the first swing stability margin of a power system from time domain simulation, IEE Proceedings on Generation, Transmission, and Distribution, vol. 143, no. 5.

[16] Horowitz, S. H. and Phadke, A. G. (2006) Third Zone Revisited, IEEE Trans. on Power Delivery, 21(1), pp. 23–29.

[17] Cohn, N. (1971) *Control of Generation and Power Flow on Interconnected Systems*, John Wiley & Sons, Ltd.

[18] Bretas, A. S. and Phadke, A. G. (2003) Artificial Neural Networks in power system restoration, IEEE Trans. on Power Delivery, vol. 18, no. 4, pp. 1181–1186.

[19] Adibi, M. M. (2000) *Power System Restoration Methodologies and Implementation Strategies*, IEEE Press, Power Engineering Series.

Appendix A

Representative system data

Computer relay design requires testing of algorithms with waveforms of currents and voltages as they are found in realistic power systems. Simulation of transients in a laboratory with physical system models or with simulation programs are useful methods of generating such test waveforms. We have collected some representative system constants for important power system elements which may be helpful in many simulation studies.

Transmission lines[1]

362 kV transmission line with horizontal conductor arrangement. The series impedance in ohms per mile at 60 Hz is given below in a lower triangular matrix for in order to save writing. The complete matrix is of course symmetric. The ground wire effect has been included in the three-by-three matrix.

$$Z_s = \begin{bmatrix} 0.2920 + j\,1.002 & & \\ 0.1727 + j\,0.4345 & 0.2359 + j\,0.9934 & \\ 0.1687 + j\,0.3549 & 0.1727 + j\,0.4345 & 0.2920 + j\,1.002 \end{bmatrix}$$

The shunt admittance matrix in micro-mhos per mile (again the lower triangular portion only) is given by

$$Y_s = \begin{bmatrix} j\,6.341 & & \\ -j\,1.115 & j\,6.571 & \\ -j\,0.333 & -j\,1.115 & j\,6.341 \end{bmatrix}$$

Computer Relaying for Power Systems 2e   by A. G. Phadke and J. S. Thorp
© 2009 John Wiley & Sons, Ltd

(b) 800 kV transmission line with horizontal configuration. The series impedance matrix is given by

$$Z_s = \begin{bmatrix} 0.1165 + j\,0.8095 & & \\ 0.0097 + j\,0.2961 & 0.1176 + j\,0.7994 & \\ 0.0094 + j\,0.2213 & 0.0097 + j\,0.2961 & 0.1165 + j\,0.8095 \end{bmatrix}$$

and the shunt admittance matrix is given by

$$Y_s = \begin{bmatrix} j\,7.125 & & \\ -j\,1.133 & j\,7.309 & \\ -j\,0.284 & -j\,1.133 & j\,7.125 \end{bmatrix}$$

To simulate transposed lines, it is sufficiently accurate to replace all the off-diagonal entries by their average value; and replace all the diagonal entries by their average. If D are the diagonal entries and M the off-diagonal entries, the positive and negative sequence values are given by $(D - M)$, while the zero sequence values are given by $(D + 2M)$.

(c) Parallel Transmission Lines. The following data are for double circuit 500 kV transmission lines. The series impedance matrix is given by

$$Z_s = \begin{bmatrix} Z_{s1} & Z_m \\ Z_m^T & Z_{s2} \end{bmatrix}$$

$$Y_s = \begin{bmatrix} Y_{s1} & Y_m \\ Y_m^T & Y_{s2} \end{bmatrix}$$

where

$$Z_{s1} = \begin{bmatrix} 0.1964 + j\,0.9566 & & \\ 0.1722 + j\,0.3833 & 0.2038 + j\,0.9477 & \\ 0.1712 + j\,0.2976 & 0.1769 + j\,0.3749 & 0.2062 + j\,0.9397 \end{bmatrix}$$

$$Z_{s2} = \begin{bmatrix} 0.2062 + j\,0.9397 & & \\ 0.1769 + j\,0.3749 & 0.2038 + j\,0.9477 & \\ 0.1712 + j\,0.2976 & 0.1722 + j\,0.3833 & 0.1964 + j\,0.9566 \end{bmatrix}$$

$$Z_m^T = \begin{bmatrix} 0.1642 + j\,0.2150 & 0.1701 + j\,0.2422 & 0.1742 + j\,0.2863 \\ 0.1600 + j\,0.1922 & 0.1657 + j\,0.2120 & 0.1769 + j\,0.3749 \\ 0.1548 + j\,0.1769 & 0.1600 + j\,0.1920 & 0.1642 + j\,0.2150 \end{bmatrix}$$

$$Y_{s1} = \begin{bmatrix} j\,6.318 & & \\ -j\,1.092 & j\,6.522 & \\ -j\,0.316 & -j\,1.092 & j\,6.388 \end{bmatrix}$$

$$Y_{s2} = \begin{bmatrix} j6.388 & & \\ -j1.070 & j6.522 & \\ -j0.316 & -j1.092 & j6.318 \end{bmatrix}$$

$$Y_m^T = \begin{bmatrix} -j0.069 & -j0.1433 & -j0.4339 \\ -j0.034 & -j0.0572 & -j0.1433 \\ -j0.025 & -j0.0344 & j0.0693 \end{bmatrix}$$

Transformers

A typical EHV auto-transformer rated at 250 MVA per phase and voltage ratings of 800/345/34.5 kV has leakage reactances of 12.2%, 57%, and 37.6% between 800/345 kV, 800/34.5 kV, and 345/34.5 kV windings respectively. The magnetizing current at rated voltage is 0.058%. The saturation characteristic is defined by two additional points: a current of 0.015% at a voltage of 80%, and a current 0.29% at a voltage of 115%. This information is generally adequate for simulating transient phenomena of interest for relaying. Note that for the leakage reactances given, the equivalent circuit has one branch with a negative reactance. If a physical model is developed, the branch will probably be set equal to zero. A digital computer simulation on the other hand can accommodate the negative value.

Generators

A modern 1300 MW cross-compound generator has x_d of 1.92, x_q of 1.85, x_d' of 0.34 and x_q' of 0.60 on its own base. The per unit rotor inertia constant is 8.1. A 100 MW hydro unit has x_d of 1.16, x_q of 0.75, x_d' of 0.40 and x_q' of 0.40 on its own base. Its per unit rotor inertia is 4.2.

It is rare that any other constants will be needed in relaying studies. The transient time constants of the generators are of the order of several seconds.

Power system

The power system used in relaying studies should contain sufficient complexity to represent conditions seen by relays in actual service. Thus, in studying transmission line relaying, each line terminal should be connected to at least few other lines. Lines of 50 to 100 miles in length are not uncommon on a high voltage network, and produce significant transient components in voltage and current waveforms which may produce problems for the relays. It may be necessary to use detailed line models (several π-sections) two or three buses removed from the transmission line terminals. Multiple-winding transformers should be modeled correctly, taking note of the proper zero and positive sequence impedances. Transformer core saturation should also be modeled realistically, remembering that the inrush and over-excitation transients are the key factors in determining the behavior of transformer relays. Models of current and voltage transformers should also be included. If digital simulation of the power system is used, care should be taken to make sure that the waveforms do not exhibit steep changes between time-steps of the simulation due to the finite interval integration algorithm used. Often low pass filtering of the output will take care of such phenomena. Averaging of successive time step outputs may be sufficient in most cases.

References

[1] Transmission Line Reference Book: 345 kV and Above, Second Edition, 1982, Electric Power Research Institute, Palo Alto, California.

Appendix B

Standard sampling rates

Note: At the time of printing the first edition of this work, these sampling rates were proposed to be a part of the standard COMTRADE. This standard has now been in force for many years, and these rates are part of that standard.

In a substation with digital protection, monitoring, and control systems, many sampling rates may be called for. Oscillography, for example, may require signals sampled at several kHz, while relaying computers may need sampling rates of the order of several hundred Hz. Also, it seems certain that in the future, as computers and Analog to Digital Converters acquire greater capabilities, the entire spectrum of sampling rates may be shifted upward. When sampling at various sampling rates is required, it seems reasonable to sample at the highest feasible sampling rate in the Data Acquisition Unit, and convert these samples to the required lower rates within the subsystem which calls for the lower rate. The highest sampling process should use an anti-aliasing filter that is commensurate with that rate. It then becomes necessary to devise algorithms which will convert these data to the lower sampling rates with *their* corresponding anti-aliasing filters. The same argument is valid when we consider that the data may be obtained at one site, and used by some user as a test case for off-line simulations or testing of algorithms. Here also, conversion to another sampling rate with accompanying anti-aliasing filtering is needed.

An IEEE standard recommends two lists of sampling rates which lead to convenient sampling rate conversion algorithms.[1] Let f_1 and f_2 be the sampling rates at which the data are obtained and at which they are desired, respectively. It is assumed that f_1 is higher than f_2. If L and M are two integers, such that

$$Lf_1 = Mf_2 = f_{LCM}$$

Computer Relaying for Power Systems 2e by A. G. Phadke and J. S. Thorp
© 2009 John Wiley & Sons, Ltd

Table B.1 Recommended sampling frequencies
corresponding to $f_{LCM} = 384 \times f_{base}$

L,M	Samples/cycle	f for 60 Hz	f for 50 Hz
1	384	23040	19200
2	192	11520	9600
3	128	7680	6400
4	96	5760	4800
6	64	3840	3200
8	48	2880	2400
12	32	1920	1600
16	24	1440	1200
24	16	960	800
32	12	720	600
48	8	480	400
64	6	360	300
96	4	240	200

Table B.2 Recommended sampling frequencies
corresponding to $f_{LCM} = 3200 \times f_{base}$

L,M	Samples/cycle	f for 60 Hz	f for 50 Hz
1	3200	192000	160000
2	1600	96000	80000
4	800	48000	40000
5	640	38400	32000
8	400	24000	20000
10	320	19200	16000
16	200	12000	10000
20	160	9600	8000
25	128	7680	6400
32	100	6000	5000
40	80	4800	4000
50	64	3840	3200
64	50	3000	2500
80	40	2400	2000
100	32	1920	1600
160	20	1200	1000
200	16	960	800
320	10	600	500
400	8	480	400
800	4	240	200

then relatively simple algorithms can be designed for the conversion process. It is of course necessary that the data obtained at f_1 and at f_2 be appropriately filtered to eliminate aliasing errors. At this time, it seems unlikely that any application of digital signal processing in a substation will call for sampling rates higher than 3600 times per cycle. Based upon this, Tables B.1 and B.2. are proposed as the standard sampling rate tables to be used in any sampling process. The DAU should use as high a sampling rate as possible from these tables. Any of the lower rates can then be obtained by simple algorithms. One such algorithm is given in Appendix C. The tables list sampling frequencies for 60 Hz and 50 Hz power systems. It should be noted that the present day relaying algorithms use only a few frequencies at the lower end of these tables. Also, most of the known relaying algorithms use frequencies from Table B.1.

References

[1] IEEE Standard Common Format for Transient Data Exchange (COMTRADE) for Power Systems. C37. 111–1991.

Appendix C

Conversion between different sampling rates

Note: This program is taken from the IEEE Standard C37.111 which is Reference 1, listed at the end of this Appendix. Fortran program is not commonly used at present, but we present it as a topic of interest, and also to give a flavor of the early work in this area. Standard subroutines and functions are now available to do the decimation process.

These FORTRAN programs have been taken from a Draft document[1], which deals with various issues of standardization in the computer based substation systems. The programs convert data between two compatible sampling rates in the sense of Appendix B, with appropriate anti-aliasing filtering provided at the lower sampling rate. The filter design is specified by its impulse response specified over one cycle. A sample program which produces such a response for a given filter transfer function is also provided.

```
C          PROGRAM CONVERT
C          CONVERTS SAMPLES TAKEN AT ONE RATE TO A SECOND
C          RATE
C          USER SUPPLIED FILTER IS IN FOR020.DAT
C          DATA IS IN FOR021.DAT
C          OUTPUT IS IN FOR025.DAT
C          NFMAX = THE MAXIMUM LENGTH OF THE FILTER
           PARAMETER NFMAX = 3600
C          3600 CORRESPONDS TO ONE CYCLE
C          LFAC = THE NUMBER OF TENTHS OF A DEGREE BETWEEN
C          SAMPLES IN INPUT
           PARAMETER LFAC = 50
```

```
C                FSAMP = THE INPUT SAMPLING FREQUENCY
                 PARAMETER FSAMP = 4320
C                NSIZE = THE MAXIMUM LENGTH OF THE INPUT DATA
C                STRING
                 PARAMETER NSIZE = 720
                 INTEGER*2 DBUF(NSIZE)
                 DIMENSION HFIL(NFMAX),DTD(NFMAX)
                 DATA N0/0/
C                GET FILTER RESPONSE
                 READ(20,*)NA,NB
                 IF(NB.LE.NFMAX) GO TO 6
                 WRITE(6,5)
5                FORMAT(3X,'DECIMATION FILTER IS TOO LONG')
                 STOP
C
6                NBF = NB/LFAC
                 IF(NB.EQ.NBF*LFAC) GO TO 10
                 WRITE(6,*)'FILTER LENGTH INDIVISIBLE BY LFAC'
                 STOP
C
10               READ(20,*)(HFIL(JJ),JJ = 1,NB)
C
C                **************************************************
C
                 WRITE(6,18)

18               FORMAT(1H$, 'ENTER TOTAL NUMBER OF SAMPLES TO BE
                   $
                 PROCESSED')
C
                 READ(21,*)(DBUF(JJ),JJ = 1,ITIME)
                 IPTR = 1
C
30               WRITE(6,35)
35               FORMAT(1H$,'ENTER THE DESIRED PROCESSING RATE')
                 READ(6,*)DRATE
                 MFAC = IFIX(FSAMP*LFAC/DRATE)
                 IF(MFAC*DRATE.EQ.FSAMP*LFAC) GO TO 40
C
                 WRITE(6,*)'RATE IS UNACHIEVABLE – TRY AGAIN'
                 GO TO 30
C
```

```
40          WRITE(6,*)'INTERPOLATION FACTOR = ',LFAC
            WRITE(6,*)'DECIMATION FACTOR = ',MFAC
C
C           ****************************************
C           DO 500 I = 1,ITIME
            DT = (I-1)/4320
            X = FLOAT(DBUF(IPTR)
            WRITE(26,*)DT,X
C
            DO 120 J = 1,NBF-1
            INDX = NBF+1-J
120         ZTD1(INDX) = ZTD1(INDX-1)
            ZTD1(1) = X
C
            N0 = N0+LFAC
            IF(N0.LT.MFAC) GO TO 500
C
            N0 = N0-MFAC
C
            ZOUT = 0.
            DO 130 J = 1,NBF
            INDX = J*LFAC-N0
130         ZOUT = ZOUT+HFIL(X)*ZTD1(J)
            ZOUT = ZOUT/FSAMP
            WRITE(25,*)DT,ZOUT
C
500         CONTINUE
            STOP
            END
            PROGRAM FIR
C           ****************************************************
C           IMPUSLE INVARIANT DESIGN FOR SECOND ORDER
C           LOW-PASS FILTER WITH REAL POLES AT – S1 AND – S2
C
C           TRANSFER FUNCTION = A*S1*S2/(S+S1)(S+S2)
C
C           SAMPLING RATE OF 216000 AT 60 HZ
C           180000 AT 50 HZ
C
C           ONE CYCLE DURATION FINITE IMPULSE RESPONSE FILTER
C           OBTAINED BY WRITING THE PARTIAL FRACTION
```

```
C           EXPANSION OF THE TRANSFER FUNCTION AND FORMING
C           THE IMPULSE RESPONSE IN THE FORM
C           H(T) = SUM(CI*EXP(-SI*T)
C           ************************************************
            DIMENSION h(3600)
            S1 = 394
            S2 = 2630
C           MAKE GAIN AT 60 HZ = 1
C           G60 = THE INVERSE OF THE 60 HZ GAIN
            G60 = (SQRT((S1**2 + (377)**2)*(S2**2 + (377)**2)))/(S1*S2)
            C1 = G60*S1*S2/(-S1+S2)
            C2 = G60*S1*S2/(S1-S2)
            WRITE(20,*)1,3600
C
            DO 100 I = 1,3600
            DT = (I-1)/216000
            H(I) = C1*EXP(-DT*S1)C2*EXP(-DT*S2)
            WRITE(20,*) H(I)
100         CONTINUE
C
            STOP
            END
```

References

[1] IEEE Standard Common Format for Transient Data Exchange (COMTRADE) for Power
     Systems. C37. 111–1991.

Appendix D

Standard for transient data exchange

Note: At the time of the first edition the IEEE Standard C37.111 was in a draft form. It has now been issued as a standard by both IEEE and IEC, and is one of the most used standards in the substation automation and monitoring systems.

This standard format is from a Draft document (now a standard)[1] which has developed and proposed several standards suitable for a computer based protection, monitoring, and control systems. Although it is likely that some such standard will be adopted by the industry, it may undergo substantial changes before it is actually accepted. One should therefore investigate the status of these proposed documents at the time they are to be used.

It is recognized that there are various sources of transient records from power systems, and many potential users of these data. For example, the data may be obtained from actual power system oscillographs, model system simulators, or digital computer simulation programs. The data could be used for post-event sequence-of-events analysis, monitoring or equipment health, in checking the performance of various protective devices, or as an aid in designing newer protection and control algorithms. It thus becomes necessary to agree upon a standard for exchanging these records between various people at various times in such a fashion that the user may be assured of finding the data in a well known and agreed upon – in other words, a standard – format. Many hardware designers would prefer to have such a standard, so that their output and input, by conforming to this standard will have universal interface capability.

The proposed standard acknowledges that one of the primary methods of exchanging data among different people would be by mailing the data in their own storage medium. Therefore, the storage medium proposed in the standard is the 5-1/4 inch floppy diskette or its equivalent (such as the newer 3-1/2 inch micro floppy diskette) prepared on the IBM or IBM compatible personal computer using PC-DOS OR

Computer Relaying for Power Systems 2e   by A. G. Phadke and J. S. Thorp
© 2009 John Wiley & Sons, Ltd

ms-dos operating systems. The files are in ASCII character format, and will be in two parts. The first part is the HEADER, and contains such textual information as the case description, signal conditioning applied to the signals, channels and the signals recorded on each channel, sampling rate, units of signals on each channel, time and date when the data were obtained, and any other information which helps to interpret the data.

The second part of the file consists of the DATA, arranged in rows of sample values taken at each sampling instant. The sample instant and sample number are identified, and each data entry is separated by a comma from the next entry. A provision is made to identify bad samples (invalid samples) with special codes which are identified in the header information. The reference cited provides an example of a transient data case stored in the standard format.

References

[1] IEEE Standard Common Format for Transient Data Exchange (COMTRADE) for Power Systems. C37. 111–1991.

Index